Lecture Notes in Mathematics 2236

More information about this series at http://www.springer.com/series/304

Petteri Harjulehto • Peter Hästö

Orlicz Spaces
and Generalized
Orlicz Spaces

 Springer

Petteri Harjulehto
Department of Mathematics and Statistics
University of Turku
Turku, Finland

Peter Hästö
Department of Mathematics and Statistics
University of Turku
Turku, Finland

ISSN 0075-8434 ISSN 1617-9692 (electronic)
Lecture Notes in Mathematics
ISBN 978-3-030-15099-0 ISBN 978-3-030-15100-3 (eBook)
https://doi.org/10.1007/978-3-030-15100-3

Mathematics Subject Classification (2010): Primary: 46E30; Secondary: 46E35, 42B25, 42B20

This Springer imprint is published by the registered company Springer Nature Switzerland AG
The registered company address is: Gewerbestrasse 11, 6330 Cham, Switzerland

Preface

Generalized Orlicz spaces have been an area of growing interest recently. In this book, we present a systematic treatment of these spaces in a general framework.

We have tried to make the book accessible to graduate students and a valuable resource for researchers. The manuscript was used in an advanced analysis course at the University of Turku in the fall of 2018, and we thank the participants of the course for their input.

We thank our friends and colleagues for their support. We are especially grateful to Ritva Hurri-Syrjänen, Jonne Juusti, Arttu Karppinen, Tetsu Shimomura, and the anonymous reviewers for their helpful comments and corrections. We also thank our collaborators on the topic of general Orlicz spaces: Debangana Baruah, David Cruz-Uribe, Rita Ferreira, Riku Klén, Mikyoung Lee, Olli Toivanen, Jihoon Ok, and Ana Ribeiro.

We also wish to express our appreciation of our colleagues whose results are presented and ask for understanding for possible lapses, omissions, and misattributions. Finally, we hope that the readers will find this book useful.

Turku, Finland   Petteri Harjulehto
Turku, Finland   Peter Hästö
February 2019

Contents

List of Symbols

φ_B^-	Essential infimum of φ in B, page 89
φ_B^+	Essential supremum of φ in B, page 41
φ_B^+	Essential supremum of φ in B, page 89
$\varphi,\ \psi$	(Generalized) Φ-function, page 14
φ^*	Conjugate Φ-function of φ, page 30
Φ_c	Set of convex Φ-functions, page 14
$\Phi_c(A,\mu)$	Set of generalized convex Φ-functions, page 37
φ^{-1}	Left-inverse of φ, page 23
Φ_s	Set of strong Φ-functions, page 14
$\Phi_s(A,\mu)$	Set of generalized strong Φ-functions, page 37
Φ_w	Set of weak Φ-functions, page 14
$\Phi_w(A,\mu)$	Set of generalized weak Φ-functions, page 37
p'	Dual exponent, $\frac{1}{p}+\frac{1}{p'}=1$, page 9
ϱ_φ	Modular function, page 48
$\mathbb{R}^n$	Euclidean, n-dimensional space, page 6
S	Simple functions, page 66
T_B	Averaging operator, page 96
t_∞	$\inf\{t\ :\ \varphi(t)=\infty\}$, page 24
T_k	Averaging operator over dyadic cubes, page 129
t_0	$\sup\{t\ :\ \varphi(t)=0\}$, page 24
$[w]_{A_p}$	Muckenhoupt "norm", page 106
$W^{k,\varphi}$	Sobolev space, page 124
$H_0^{k,\varphi}$	Sobolev space with zero boundary values, page 127
$X \cap Y$	Intersection of normed spaces, page 7
$X + Y$	Sum of normed spaces, page 7
X^*	Dual space of X, page 8
$[0,\infty]$	Compactification of $[0,\infty)$, page 6

Chapter 1
Introduction

1.1 Rationale and Philosophy of This Book

Some years ago, we wrote a book on variable exponent spaces with Lars Diening
and Michael Růžička [34]. One aim of that book was to avoid unnecessarily
assuming that the variable exponent is bounded. In particular, from the classical
theory it is clear that the boundedness of the Hardy–Littlewood maximal operator
should not be affected by large values of the exponent, as long as the exponent is
bounded away from 1. Consequently, we refined the trick, called the *key estimate*,
introduced a decade earlier by Diening [32]. This resulted in Theorem 4.2.4 in our
earlier book.

The purpose of the current book is to take this philosophy even further in the
context of Orlicz and generalized Orlicz spaces (also known as Musielak–Orlicz
spaces). For researchers acquainted with our previous book [34], the approach
adopted here might seem familiar, since we have tried to structure our arguments
to parallel as closely as possible the variable exponent case. Thus we hope that
our tools and exposition will aid in generalizing further results from the variable
exponent setting to the generalized Orlicz setting.

Our approach differs from that in standard references such as Krasnosel'skiĭ–
Rutickiĭ [75] and Rao–Ren [117]. The classical Orlicz approach emphasizes the
exact form of the norm and distinguishes between equivalent norms of the spaces.
In contrast, we emphasize choosing the representative of an equivalence class
of Φ-functions in appropriate ways so as to simplify proofs, see for example
Corollary 3.6.7 or Theorem 5.5.1. The origin of this approach is connected to the
generalized Φ-function situation (see Example 2.5.3), but there are ramifications
also in the non-generalized case, as demonstrated by the next simple example.

The classical approach is based on convex Φ-functions such as $t \mapsto t^{\frac{11}{10}}$ or
$t \mapsto t^3$. However, the minimum of Φ-functions need not be convex, as is the case
for $t \mapsto \min\{t^{\frac{11}{10}}, t^3\}$, see Fig. 1.1. But in the generalized Orlicz case it is useful to

© Springer Nature Switzerland AG 2019
P. Harjulehto, P. Hästö, *Orlicz Spaces and Generalized Orlicz Spaces*,
Lecture Notes in Mathematics 2236,
https://doi.org/10.1007/978-3-030-15100-3_1

Fig. 1.1 Function
$t \mapsto \min\{t^{\frac{11}{10}}, t^3\}$ is not
convex

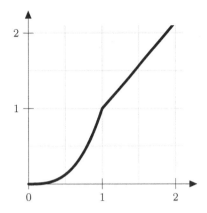

deal with the minima of nearby functions like $\inf_{x \in B} \varphi(x, t)$ (see the proof of the key estimate, Theorem 4.3.2). Furthermore, the class of convex Φ-functions is not closed under equivalence of Φ-functions, see Example 2.1.4.

The basic problem is that convexity can be destroyed at a single point, even though by-and-large the function is quite convex-like. We formalize this notion by requiring that the function

$$t \mapsto \frac{\varphi(t)}{t}$$

is almost increasing on $(0, \infty)$. It turns out that this is a sufficient property to get a quasi-norm $\| \cdot \|_\varphi$ by the usual Luxemburg procedure (see Lemma 3.2.2). Furthermore, this property is much more resilient to perturbations of the Φ-function. We call functions satisfying this condition *weak Φ-functions*. We were inspired to pursue this line of research by the work by our Japanese colleagues (cf. Sect. 7.3), who also relinquished the convexity assumption.

It turns out that this more flexible notion of Φ-functions, coupled with the emphasis on properties invariant under equivalence of Φ-functions allows us to prove several basic properties in settings more general than previously known.

In Orlicz spaces, one often restricts one's attention to so-called N-functions, thereby excluding the cases L^1 and L^∞. As was mentioned, in [34] we tried to avoid this assumption. However the tools used did not completely allow for this in [34]. Chapter 2 of our previous book was written in the context of generalized Orlicz spaces, however, some key results such as the estimate

$$(\varphi^*)^{-1}(t) \approx \frac{t}{\varphi^{-1}(t)}$$

[34, (2.6.12)], where $\varphi^*(t) := \sup_{s \geqslant 0}(st - \varphi(s))$ is the conjugate function, were proven only in the restricted setting of N-functions. In Theorem 2.4.8 of the prosent book, we are able to generalize this result to all weak Φ-functions. The technique of that proof is also note-worthy. The first lines of that proof are as follows:

> By Theorem 2.3.6 and Lemma 2.5.8(c) the claim is invariant under the equivalence of Φ-functions, so by Theorem 2.5.10 we may assume that $\varphi \in \Phi_s$.

In other words, we have established that the parts of the claim are invariant under equivalence, which allows us to switch to a more regular, so-called strong Φ-function, which is convex and continuous. Analogous sentences occur in many other proofs, as well.

This technique requires some initial investment to obtain the "equivalence-machine" going. These investments are the content of Chap. 2. We believe that Chap. 2 will pay dividend not only in the remainder of the book, but also for other researchers who adopt the methods developed here. For this reason we think that the book will be useful also for researchers interested in the Orlicz case only.

For generalized Orlicz spaces, we can give additional motivation for this book. The standard reference in the field is the book by Julian Musielak from 1983 [96]. The results and presentation are therefore partially out-dated. Furthermore, Musielak's book did not touch upon harmonic analysis in generalized Orlicz spaces as this theory was developed only in the last five years, starting with research by Maeda, Mizuta, Ohno, and Shimomura [79]. However, while this book was being written, also Lang and Méndez [76] have written a book on Musielak–Orlicz spaces, but their perspective is more classical than ours and does not include harmonic analysis. We want to mentioned two themes that are not covered by this book. Musielak–Orlicz Hardy spaces have been studied by Yang, Liang, and Ky in the book [126] and the survey [21] by Chlebicka covers differential equations in Musielak–Orlicz spaces.

Many of the results in this book have appeared in some form in our articles [11, 29, 42, 52–62], written in collaboration with Baruah, Cruz-Uribe, Ferreira, Karppinen, Klén, Lee, Toivanen, Ok and Ribeiro. However, almost all results have been improved in this book by the generalization of the assumptions and/or the refinement of the proofs (e.g. Proposition 2.4.9 and Theorem 2.4.8). Furthermore, we have, obviously, unified the assumptions and notation. Some results are presented here for the first time (e.g. Lemmas 2.3.11, 4.2.7 and 5.2.3, Propositions 2.3.13 and 7.2.4 and Theorems 3.5.2, 4.4.7 and 6.4.7).

1.2 History of Non-standard Growth Phenomena

Variable exponent Lebesgue spaces appeared in the literature for the first time already in a 1931 article by W. Orlicz [113]. In this article the following question is considered: let (p_i), with $p_i > 1$, and (x_i), $x_i \geqslant 0$ be real-valued sequences. What is the necessary and sufficient condition on (y_i) for $\sum_i x_i y_i$ to converge whenever $\sum_i x_i^{p_i}$ converges? It turns out that the answer is that $\sum_i (\lambda y_i)^{p_i'}$ should converge for some $\lambda > 0$ and $p_i' = \frac{p_i}{p_i-1}$. This is essentially Hölder's inequality in the space

$\ell^{p(\cdot)}$. Orlicz also considered the variable exponent function space $L^{p(\cdot)}$ on the real line, and proved the Hölder inequality in this setting. Variable exponent spaces have been studied in more than a thousand papers in the past 15 years so we only cite a few monographs on the topic which can be consulted for additional references [18, 27, 34, 38, 72, 73, 115]. All central results of harmonic analysis in variable exponent spaces are covered as special cases of the results in this book, see Sect. 7.1.

After this one paper [113], Orlicz abandoned the study of variable exponent spaces, to concentrate on the theory of the function spaces that now bear his name (but see also [97]). In the theory of Orlicz spaces, one defines the space L^{φ} in an open set $\Omega \subset \mathbb{R}^n$ to consist of those measurable functions $u \colon \Omega \to \mathbb{R}$ for which

$$\varrho(\lambda u) = \int_{\Omega} \varphi(\lambda |u(x)|)\, dx < \infty$$

for some $\lambda > 0$ (φ has to satisfy certain conditions, see Definition 2.1.3). Abstracting certain central properties of ϱ, one is led to a more general class of so-called modular function spaces which were first systematically studied by H. Nakano [98, 99]. In the appendix [p. 284] of the first of his books, Nakano mentions explicitly variable exponent Lebesgue spaces as an example of the more general spaces he considers. The duality property is again observed.

Following the work of Nakano, modular spaces were investigated by several people, most importantly by the groups at Sapporo (Japan), Voronezh (USSR), and Leiden (Netherlands). Somewhat later, a more explicit version of these spaces, modular function spaces, were investigated by Polish mathematicians, for instance by H. Hudzik, A. Kamińska and J. Musielak. For a comprehensive presentation of modular function spaces and generalized Orlicz spaces, see the monograph [96] by Musielak.

Harmonic analysis in generalized Orlicz spaces has only recently been studied. In 2005, L. Diening [33] investigated the boundedness of the maximal operator on L^{φ} and gave abstract conditions on the Φ-function for the boundedness to hold. In the variable exponent case it led to the result that the maximal operator is bounded on $L^{p(\cdot)}(\mathbb{R}^n)$ if and only if it is bounded on $L^{p'(\cdot)}(\mathbb{R}^n)$ (provided $1 < p^- \leqslant p^+ < \infty$). Unfortunately, this result has still not been successfully extended to generalized Orlicz spaces.

A different route was taken in 2013 by F.-Y. Maeda, Y. Mizuta, T. Ohno, Y. Sawano and T. Shimomura (e.g. [79–87, 94, 102–107, 123]): they proposed some concrete conditions on φ which are sufficient for the boundedness of the maximal operator, but not necessary. This is similar to the path followed for variable exponent spaces, where the sufficient log-Hölder condition is used in the most studies. This is also the route taken in this book, although our sufficient conditions are more general (cf. Sect. 7.3).

A related field of research is partial differential equations with non-standard growth. Consider minimizers of the non-autonomous quasilinear elliptic problem

$$\min_{u \in W^{1,1}(\Omega)} \int_\Omega F(x, \nabla u) \, dx. \tag{1.2.1}$$

If $F(x, t) \approx |t|^p$ (in some suitable sense) Marcellini [89, 90] called this the *standard growth* case. He studied the more general (p, q)-growth case $t^p \lesssim F(x, t) \lesssim t^q + 1$, $q > p > 1$, and found that results from the standard growth case generalize when the ratio $\frac{q}{p}$ is sufficiently close to 1 with different bounds for different properties. For instance, the minimizers in $W^{1,q}$ are Lipschitz if $\frac{q}{p} \leqslant \frac{n}{n-2}$, where $n > 2$ is the dimension [90] and the minimizer has higher integrability if $\frac{q}{p} \leqslant 1 + \frac{\alpha}{n}$ where F is α-Hölder continuous [39]. Also Zhikov [129, 130] studied several special cases of non-standard growth in the 1980's, including the variable exponent case $\varphi(x, t) = t^{p(x)}$ and the double phase case $\varphi(x, t) = t^p + a(x)t^q$. The latter has been recently studied by P. Baroni, M. Colombo and G. Mingione [7–9, 25, 26], and it is interesting to note that their conditions are also recovered as special cases of ours, see Sect. 7.2. In [57, 61] we have extended several of their results on regularity of minimizers to the generalized Orlicz case, using the results of this book, see also [78, 125].

1.3 Notation and Background

In this section we clarify the basic notation used in this book. Moreover we give precise formulations of some basic results which are frequently used.

We use the symbol $:=$ to define the left-hand side by the right-hand side. For constants we use the letters $c, c_1, c_2, C, C_1, C_2, \ldots$, or other letters specifically mentioned to be constants. The symbol c without index stands for a generic constant which may vary between appearances. We use $f \approx g$ and $f \lesssim g$ if there exist constants $c_1, c_2 > 0$ such that $c_1 f \leqslant g \leqslant c_2 f$ and $f \leqslant c_2 g$, respectively. For sets A and B the notation $A \subset B$ includes also the case $A = B$. We use the terms *increasing* and *strictly increasing* when $s < t$ implies that $f(s) \leqslant f(t)$ and $f(s) < f(t)$, respectively, and analogously *decreasing* and *strictly decreasing*. The notation $a \to b^-$ means that a tends to b from below, i.e. $a \leqslant b$ and $a \to b$; the notation $a \to b^+$ is defined analogously. The average integral of f over the set A, $\mu(A) > 0$, is denoted by

$$\fint_A f \, d\mu := \frac{1}{\mu(A)} \int_A f \, d\mu.$$

The *characteristic function* of a set E is denoted by χ_E.

By $\mathbb{R}^n$ we denote the n-dimensional Euclidean space, and $n \in \mathbb{N}$ always stands for the dimension of the space. By U and V we denote open sets and by F closed sets of the topological space under consideration, usually $\mathbb{R}^n$. A compact set will usually be denoted by K. We use the notation $A \subset\subset E$ if the closure $\overline{A}$ is compact and $\overline{A} \subset E$. By Ω we always denote an open subset of $\mathbb{R}^n$. If the set has additional properties it will be stated explicitly. A *domain* $\Omega \subset \mathbb{R}^n$ is a connected open set. We will also use domains with specific conditions on the boundary, such as John domains (cf. Definition 6.2.1). The set $[0, \infty]$ is the compactification of $[0, \infty)$ with the topology generated by the chordal metric: $q(x, y) = \frac{|x-y|}{(1+|x|)(1+|y|)}$ for $x, y < \infty$ and $q(x, \infty) = \frac{1}{1+|x|}$. Moreover, we write $\overline{\mathbb{R}} = \mathbb{R} \cup \{\pm\infty\}$.

Open balls will be denoted by B. The open ball with radius r and center $x \in \mathbb{R}^n$ will be denoted by $B(x, r)$. We usually denote open cubes in $\mathbb{R}^n$ by Q, and by a cube we always mean a non-degenerate cube with faces parallel to the coordinate axes. By $Q(x, r)$ we denote a cube with side length r and center at $x \in \mathbb{R}^n$. A cube is *dyadic* if it is of the form $Q(x, 2^k)$ for some $x \in 2^k \mathbb{Z}^n$ and $k \in \mathbb{Z}$. For a ball B we will denote the ball with α times the radius and the same center by αB. For $a, b \in \mathbb{R}^n$ we use (a, b) and $[a, b]$ to denote the *open and closed segment*, respectively, between a and b.

Convex Functions

In this section we present some basic properties of convex functions without proofs. Proofs can be found for example in "Convex Functions and their Applications" by Niculescu and Persson [101].

In this section $I \subset \mathbb{R}$ is an interval.

Definition 1.3.1 A function $\varphi : I \to \overline{\mathbb{R}}$ is *convex* if

$$\varphi(\theta t + (1 - \theta)s) \leqslant \theta \varphi(t) + (1 - \theta)\varphi(s)$$

for every $t, s \in I$ and $\theta \in [0, 1]$.

Assume that $\varphi : [0, \infty) \to [0, \infty]$ is convex and $\varphi(0) = 0$. Choosing $s = 0$ in the definition and $\theta = \lambda$ or $\theta = \frac{1}{\lambda}$, we obtain

$$\varphi(\lambda t) \leqslant \lambda \varphi(t) \qquad \text{for } \lambda \in [0, 1],$$

$$\varphi(\lambda t) \geqslant \lambda \varphi(t) \qquad \text{for } \lambda \geqslant 1.$$

We quote Theorems 1.1.4, 1.3.3 and 1.8.1 from [101]:

Theorem 1.3.2 *Let* $\varphi : I \to \mathbb{R}$ *be a continuous function. Then* φ *is convex if and only if* φ *is mid-point convex, that is,*

$$\varphi\left(\frac{s+t}{2}\right) \leqslant \frac{\varphi(s) + \varphi(t)}{2}$$

for all $s, t \in I$.

The *left-* and *right-derivative* of f and x is defined by the limit

$$\lim_{h \to 0^-} \frac{f(x+h) - f(x)}{h} \quad \text{and} \quad \lim_{h \to 0^+} \frac{f(x+h) - f(x)}{h}, \quad \text{respectively.}$$

Theorem 1.3.3 *Let* $\varphi : I \to \mathbb{R}$ *be a convex function. Then* φ *is continuous on the interior of* I *and has finite left and right derivatives at each point of the interior. Moreover, the left and right derivatives are increasing.*

Theorem 1.3.4 (Jensen's Inequality) *Let* $\varphi : [0, \infty) \to \mathbb{R}$ *be a convex function and* $f \in L^1(a, b)$. *Then*

$$\varphi\left(\fint_a^b |f(x)| \, dx\right) \leqslant \fint_a^b \varphi(|f(x)|) \, dx.$$

Functional Analysis

Let X be a real vector space. We say that function $\| \cdot \|$ from X to $[0, \infty]$ is a *quasinorm* if:

(a) $\|f\| = 0$ if and only if $f = 0$.
(b) $\|af\| = |a| \|f\|$ for all $f \in X$ and $a \in \mathbb{R}$.
(c) There exists $Q > 0$ such that $\|f + g\| \leqslant Q(\|f\| + \|g\|)$ for all $f, g \in X$.

If $Q = 1$ in (c), then $\| \cdot \|$ is called a *norm*. A *(quasi-)Banach space* $(X, \|\cdot\|_X)$ is a (quasi)normed vector space which is complete with respect the (quasi)norm $\| \cdot \|_X$.

Let X and Y be quasinormed vector spaces. The mapping $F : X \to Y$ is *bounded* if $\|F(x)\|_Y \leqslant C \|x\|_X$ for all $x \in X$. A linear mapping is bounded if and only if it is continuous. It is an *isomorphism* if F and F^{-1} are bijective, linear and bounded.

Suppose the quasinormed spaces X and Y are subsets of a vector space Z. Then the intersection $X \cap Y$ equipped with the quasinorm $\|z\|_{X \cap Y} = \max\{\|z\|_X, \|z\|_Y\}$ and the sum $X + Y := \{x + y : x \in X, y \in Y\}$ equipped with the quasinorm

$$\|z\|_{X+Y} = \inf\{\|x\|_X + \|y\|_Y : x \in X, y \in Y, z = x + y\}$$

are quasinormed spaces. If X and Y are quasi-Banach spaces, then $X \cap Y$ and $X + Y$ are quasi-Banach spaces as well.

The *dual space* X^* of a (quasi-)Banach space X consists of all bounded, linear functionals $F \colon X \to \mathbb{R}$. The *duality pairing* between X^* and X is defined by $\langle F, x \rangle_{X^*, X} = \langle F, x \rangle := F(x)$ for $F \in X^*$, $x \in X$. The dual space is equipped with the *dual quasinorm* $\|F\|_{X^*} := \sup_{\|x\|_X \leqslant 1} \langle F, x \rangle$, which makes X^* a quasi-Banach space.

A space is called *separable* if it contains a dense, countable subset. We denote the bidual space by $X^{**} := (X^*)^*$. A quasi-Banach space X is called *reflexive* if the *natural injection* $\iota \colon X \to X^{**}$, given by $\langle \iota x, F \rangle_{X^{**}, X^*} := \langle F, x \rangle_{X^*, X}$, is surjective. A norm $\| \cdot \|$ on a space X is called *uniformly convex* if for every $\varepsilon > 0$ there exists $\delta(\varepsilon) > 0$ such that for all $x, y \in X$ satisfying $\|x\|, \|y\| \leqslant 1$, the inequality $\|x - y\| > \varepsilon$ implies $\|\frac{x+y}{2}\| < 1 - \delta(\varepsilon)$. A quasi-Banach space X is called *uniformly convex*, if there exists a uniformly convex norm $\| \cdot \|'$, which is equivalent to the original norm of X. These properties are inherited to closed linear subspaces. More precisely we have (cf. [2, Chapter I]):

Proposition 1.3.5 *Let X be a Banach space and let $Y \subset X$ be closed. Then:*

(i) *Y is a Banach space.*
(ii) *If X is reflexive, then Y is reflexive.*
(iii) *If X is separable, then Y is separable.*
(iv) *If X is uniformly convex, then X is reflexive.*
(v) *If X is uniformly convex, then Y is uniformly convex.*

We say that a (quasi-)Banach space X is *continuously embedded* into a (quasi-)Banach space Y, $X \hookrightarrow Y$, if $X \subset Y$ and there exists a constant $c > 0$ such that $\|x\|_Y \leqslant c \|x\|_X$ for all $x \in X$. The embedding of X into Y is called *compact*, $X \hookrightarrow\hookrightarrow Y$, if $X \hookrightarrow Y$ and bounded sets in X are precompact in Y. A sequence $(x_k)_{k \in \mathbb{N}} \subset X$ is called *(strongly) convergent* to $x \in X$, if $\lim_{k \to \infty} \|x_k - x\|_X = 0$. It is called *weakly convergent* if $\lim_{k \to \infty} \langle F, x_k \rangle = 0$ for all $F \in X^*$.

Let $(X, \| \cdot \|_X)$ be a Banach space and $A \subset X$ a set. The *closure of A with respect to the norm* $\| \cdot \|_X$, $\overline{A}^{\| \cdot \|_X}$, is the smallest closed set Y that contains A. The closure of a set A is denoted by $\overline{A}$ when the space is clear from the context.

Measure and Integration

We denote by (A, Σ, μ) a *measure space*. By a *measure* we mean a set function $\mu \colon \Sigma \to [0, \infty]$ that satisfies $\mu(\emptyset) = 0$ and that is countably additive,

$$\mu \left(\bigcup_{i=1}^{\infty} A_i \right) = \sum_{i=1}^{\infty} \mu(A_i)$$

where $A_i \in \Sigma$ are pairwise disjoint. We say that the measure μ is *complete*, if $E \in \Sigma$, $F \subset E$ and $\mu(E) = 0$ imply that $F \in \Sigma$. The measure μ is *σ-finite* if A is a countable union of sets $A_i \in \Sigma$ with $\mu(A_i) < \infty$. We always assume that μ is a σ-finite, complete measure on Σ with $\mu(A) > 0$. If there is no danger of confusion we omit Σ from the notation. We always assume that our measure is not identically zero.

We use the usual convention of identifying two μ-measurable functions on A if they agree *almost everywhere*, i.e. if they agree up to a set of μ-measure zero. We denote by $L^0(A, \mu)$ the set of measurable functions $f : A \to \overline{\mathbb{R}}$. The Lebesgue integral of a measurable function $f: A \to \mathbb{R}$ is defined in the standard way (cf. [121, Chapter 1]) and denoted by $\int_A f \, d\mu$. If there is no danger of confusion we will write "measurable" instead of "μ-measurable", "almost everywhere" instead of "μ-almost everywhere", etc.

The most prominent example for our purposes are: $A = \Omega$ is an open subset of $\mathbb{R}^n$, μ is the n-dimensional Lebesgue measure, and Σ is the σ-algebra of Lebesgue-measurable subsets of Ω; or $A = \mathbb{Z}^n$, μ is the counting measure, and Σ is the power set of $\mathbb{Z}^n$. In the former case the Lebesgue integral of a function f will be denoted by $\int_\Omega f \, dx$ and the measure of a measurable subset $E \subset \Omega$ will be denoted by $|E|$. The classical *sequence spaces* $l^s(\mathbb{Z}^n)$, $s \in [1, \infty]$, are defined as the Lebesgue spaces $L^s(\mathbb{Z}^n, \mu)$ with μ being the counting measure. Thus many results of this book hold also for this case.

In the case of a measure space (A, μ) we denote by $L^s(A, \mu)$, $s \in [1, \infty]$, the classical *Lebesgue space* (cf. [2]). If there is no danger of confusion we omit the set A and the measure μ and abbreviate L^s. For an exponent $p \in [1, \infty]$ the *dual* or *conjugate exponent* p' is defined by $\frac{1}{p} + \frac{1}{p'} = 1$, with the usual convention $\frac{1}{\infty} = 0$. In most cases we do not distinguish in the notation of the function spaces and the corresponding norms between scalar or vector-valued functions.

By $L^1_{\text{loc}}(\Omega)$ we denote the space of all locally integrable functions f, i.e. $f \in L^1(U)$ for all open $U \subset\subset \Omega$. A function $f: \Omega \to \mathbb{K}$ has *compact support* if there exists a compact set $K \subset \Omega$ such that $f = f\chi_K$ where χ_K is the characteristic function of K.

We denote by $C(\overline{\Omega})$ the space of uniformly continuous functions equipped with the supremum norm $\|f\|_\infty = \sup_{x \in \overline{\Omega}} |f(x)|$. By $C^k(\overline{\Omega})$, $k \in \mathbb{N}$, we denote the space of functions f, such that the partial derivatives $\partial_\alpha f \in C(\overline{\Omega})$ for all $|\alpha| \leqslant k$. The space is equipped with the norm $\sup_{|\alpha| \leqslant k} \|\partial_\alpha f\|_\infty$. The set of *smooth functions* in Ω is denoted by $C^\infty(\Omega)$—it consists of functions which are continuously differentiable arbitrarily many times. The set $C_0^\infty(\Omega)$ is the subset of $C^\infty(\Omega)$ of functions which have compact support.

A *standard mollifier* is a non-negative, radially symmetric and radially decreasing function $\psi \in C_0^\infty(B(0, 1))$ with $\int_{B(0,1)} \psi \, dx = 1$. For standard mollifiers we often use the L^1-scaling $\psi_\varepsilon(\xi) := \varepsilon^{-n} \psi(\xi/\varepsilon)$.

A sequence (f_i) of μ-measurable functions is said to *converge in measure* to the function f if for every $\varepsilon > 0$ there exists a natural number N such that

$$\mu(\{x \in A : |f_i(x) - f(x)| \geqslant \varepsilon\}) \leqslant \varepsilon$$

for all $i \geqslant N$. For $\mu(A) < \infty$, it is well known that if $f_i \to f$ μ-almost everywhere, then $f_i \to f$ in measure; on the other hand $\lim_{i \to \infty} \int_A |f_i - f|\,d\mu = 0$ does not imply convergence in measure unless we pass to an appropriate subsequence. The following result is Theorem 3.12 in [121].

Theorem 1.3.6 *Let (A, Σ, μ) be a σ-finite, complete measure space. Assume that f and (f_i) belong to $L^1(A, \mu)$, $\mu(A) < \infty$ and*

$$\lim_{i \to \infty} \int_A |f_i - f|\,d\mu = 0.$$

Then there exists a subsequence (f_{i_k}) such that $f_{i_k} \to f$ μ-almost everywhere.

We write $a_k \nearrow a$ if (a_k) is a sequence increasing to a. We frequently use the following convergence results, which can be found e.g. in [2, 50]. The three properties are called *Fatou's lemma, monotone convergence* and *dominated convergence*, respectively.

Theorem 1.3.7 *Let (A, Σ, μ) be a σ-finite, complete measure space. Let (f_k) be a sequence of μ-measurable functions.*

(a) *If there exists $g \in L^1(A, \mu)$ with $f_k \geqslant g$ μ-almost everywhere for every $k \in \mathbb{N}$, then*

$$\int_A \liminf_{k \to \infty} f_k\,d\mu \leqslant \liminf_{k \to \infty} \int_A f_k\,d\mu.$$

(b) *If $f_k \nearrow f$ μ-almost everywhere and $\int_A f_1\,d\mu > -\infty$, then*

$$\lim_{k \to \infty} \int_A f_k\,d\mu = \int_A f\,d\mu.$$

(c) *If $f_k \to f$ μ-almost everywhere and exists $h \in L^1(A, \mu)$ such that $|f_k| \leqslant h$ μ-almost everywhere for all $k \in \mathbb{N}$, then $f \in L^1(A, \mu)$ and*

$$\lim_{k \to \infty} \int_A f_k\,d\mu = \int_A f\,d\mu.$$

1.4 Tools Missing in Generalized Orlicz Spaces

We conclude the introduction by pointing out some tools from the classical theory that do not work in the general setting that we are considering. Let us compare four different type of spaces: classical Lebesgue spaces L^p, Orlicz spaces, variable exponent Lebesgue spaces $L^{p(\cdot)}$ and generalized Orlicz spaces. Naturally, L^p-spaces are Orlicz spaces and $L^{p(\cdot)}$-spaces, and Orlicz and $L^{p(\cdot)}$-spaces are

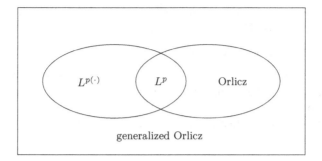

Fig. 1.2 Relation between different classes of spaces

generalized Orlicz spaces. Orlicz spaces and $L^{p(\cdot)}$-spaces have different nature, and neither of them is a subset of the other (see Fig. 1.2).

Let us list some techniques from the classical case which do not work in $L^{p(\cdot)}$-spaces and some additional ones that do not work in the generalized Orlicz case. Orlicz spaces are similar to L^p-spaces in many regards, but some differences exist.

– Exponents cannot be moved outside the Φ-function, i.e. $\varphi(t^\gamma) \neq \varphi(t)^\gamma$ in general.
– The formula $\varphi^{-1}\left(\int_\Omega \varphi(|f|)\,dx\right)$ does not define a norm.

Techniques which do not work in $L^{p(\cdot)}$-spaces (from [34, pp. 9–10]):

– The space $L^{p(\cdot)}$ is not rearrangement invariant; the translation operator $T_h : L^{p(\cdot)} \to L^{p(\cdot)}$, $T_h f(x) := f(x + h)$ is not bounded; Young's convolution inequality $\|f * g\|_{p(\cdot)} \leqslant c\,\|f\|_1\|g\|_{p(\cdot)}$ does not hold [34, Section 3.6].
– The formula

$$\int_\Omega |f(x)|^p\,dx = p \int_0^\infty t^{p-1}\big|\{x \in \Omega : |f(x)| > t\}\big|\,dt$$

has no variable exponent analogue.
– Maximal, Poincaré, Sobolev, etc., inequalities do not hold in a modular form. For instance, A. Lerner showed that the inequality

$$\int_{\mathbb{R}^n} |Mf|^{p(x)}\,dx \leqslant c \int_{\mathbb{R}^n} |f|^{p(x)}\,dx$$

holds if and only if $p \in (1, \infty)$ is constant [77, Theorem 1.1]. For the Poincaré inequality see [34, Example 8.2.7] and the discussion after it.
– Interpolation is not so useful, since variable exponent spaces never result as an interpolant of constant exponent spaces (see Sect. 5.5).

- Solutions of the $p(\cdot)$-Laplace equation are not scalable, i.e. λu need not be a solution even if u is [34, Example 13.1.9].

New obstructions in generalized Orlicz spaces:

- We cannot estimate $\varphi(x,t) \lesssim \varphi(y,t)^{1+\varepsilon} + 1$ even when $|x - y|$ is small, because of lack of polynomial growth. This complicates e.g. the use of higher integrability in PDE proofs.
- It is not always the case that $\chi_E \in L^\varphi(\Omega)$ when $|E| < \infty$.

Chapter 2
Φ-Functions

As mentioned in the introduction, we approach Φ-functions and Orlicz spaces by slightly more robust properties than the commonly used convexity. Instead, we will study "almost convex" Φ-functions and show that essentially all essential properties hold also in this case. Furthermore, this approach allows us much more room for manoeuvring and modifying the Φ-functions than the classical setup. This is the content of Sects. 2.1 and 2.2. Two central tools when working with a Φ-function φ are its inverse φ^{-1} and its conjugate φ^*, which are the topics of Sects. 2.3 and 2.4, respectively. Finally, in the concluding section, Sect. 2.5, we extend the results from the earlier sections to the generalized Φ-function $\varphi(x, t)$ depending also on the space variable x.

2.1 Equivalent Φ-Functions

In this section we set up the main element of our study, the Φ-function. We start by defining the properties of almost increasing and almost decreasing, which will be used throughout the book. Finally, we consider two notions of the equivalence of Φ-functions and prove relations between them.

Definition 2.1.1 A function $g : (0, \infty) \to \mathbb{R}$ is *almost increasing* if there exists a constant $a \geqslant 1$ such that $g(s) \leqslant ag(t)$ for all $0 < s < t$. *Almost decreasing* is defined analogously.

Increasing and decreasing functions are included in the previous definition as the special case $a = 1$.

© Springer Nature Switzerland AG 2019
P. Harjulehto, P. Hästö, *Orlicz Spaces and Generalized Orlicz Spaces*,
Lecture Notes in Mathematics 2236,
https://doi.org/10.1007/978-3-030-15100-3_2

Definition 2.1.2 Let $f : (0, \infty) \to \mathbb{R}$ and $p, q > 0$. We say that f satisfies

$(\text{Inc})_p$ if $\frac{f(t)}{t^p}$ is increasing;

$(\text{aInc})_p$ if $\frac{f(t)}{t^p}$ is almost increasing;

$(\text{Dec})_q$ if $\frac{f(t)}{t^q}$ is decreasing;

$(\text{aDec})_q$ if $\frac{f(t)}{t^q}$ is almost decreasing.

We say that f satisfies (aInc), (Inc), (aDec) or (Dec) if there exist $p > 1$ or $q < \infty$ such that f satisfies $(\text{aInc})_p$, $(\text{Inc})_p$, $(\text{aDec})_q$ or $(\text{Dec})_q$, respectively.

Suppose that φ satisfies $(\text{aInc})_{p_1}$. Then it satisfies $(\text{aInc})_{p_2}$ for $p_2 < p_1$ and it does not satisfy $(\text{aDec})_q$ for $q < p_1$. Likewise, if φ satisfies $(\text{aDec})_{q_1}$, then it satisfies $(\text{aDec})_{q_2}$ for $q_2 > q_1$ and it does not satisfy $(\text{aInc})_p$ for $p > q_1$.

Definition 2.1.3 Let $\varphi \colon [0, \infty) \to [0, \infty]$ be increasing with $\varphi(0) = 0$, $\lim_{t \to 0^+} \varphi(t) = 0$ and $\lim_{t \to \infty} \varphi(t) = \infty$. Such φ is called a Φ-*prefunction*. We say that a Φ-prefunction φ is a

– (*weak*) Φ-*function* if it satisfies $(\text{aInc})_1$ on $(0, \infty)$;
– *convex* Φ-*function* if it is left-continuous and convex;
– *strong* Φ-*function* if it is continuous in the topology of $[0, \infty]$ and convex.

The sets of weak, convex and strong Φ-functions are denoted by Φ_w, Φ_c and Φ_s, respectively.

Note that when we speak about Φ-functions, we mean the weak Φ-functions of the previous definition, whereas many other authors use this term for convex Φ-functions, possibly with additional assumptions as well. We mention the epithet "weak" only when special emphasis is needed. Note also that one of the improvements in this book is to remove the requirement of left-continuity from weak Φ-functions, whereas in some of our previous works it was included.

Continuity in the topology of $C([0, \infty); [0, \infty])$ means that

$$\lim_{s \to t} f(s) = f(t)$$

for every point $t \in [0, \infty)$ regardless of whether $f(t)$ is finite or infinite.

If φ is convex and $\varphi(0) = 0$, then we obtain for $0 < s < t$ that

$$\varphi(s) = \varphi\left(\frac{s}{t} t + 0\right) \leqslant \frac{s}{t} \varphi(t) + \left(1 - \frac{s}{t}\right) \varphi(0) = \frac{s}{t} \varphi(t), \tag{2.1.1}$$

i.e. $(\text{Inc})_1$ holds. Therefore, it follows from the definition that $\Phi_s \subset \Phi_c \subset \Phi_w$. As a convex function, φ is continuous in $[0, \infty)$ if and only if it is finite on $[0, \infty)$. Let us show that if $\varphi \in \Phi_c$ satisfies (aDec), then $\varphi \in \Phi_s$. Since $\lim_{s \to 0^+} \varphi(s) = 0$, we can find $s > 0$ with $\varphi(s) < \infty$. By (aDec) we obtain $\varphi(t) \leqslant a \frac{t^q}{s^q} \varphi(s)$ for $t > s$. Hence $\varphi < \infty$ in $[0, \infty)$ and thus it is continuous.

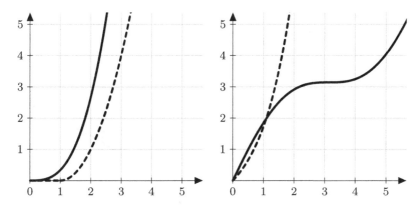

Fig. 2.1 Functions from Example 2.1.4. Left: φ^3 (solid) and $\varphi_{\max}$ (dashed). Right: $\varphi_{\sin}$ (solid) and $\varphi_{\exp}$ (dashed)

Example 2.1.4 We consider several examples to motivate and clarify our definition and the choices behind it. In order to collect the interesting observations in one place, we use here also some concepts from later in the section.

For $t \geqslant 0$, we define (see Fig. 2.1)

$$\varphi^p(t) := \tfrac{1}{p} t^p, \quad p \in (0, \infty),$$

$$\varphi_{\max}(t) := (\max\{0, t - 1\})^2,$$

$$\varphi_{\sin}(t) := t + \sin(t),$$

$$\varphi_{\exp}(t) := e^t - 1,$$

$$\varphi^\infty(t) := \infty \chi_{(1,\infty)}(t),$$

$$\varphi^{\infty,2}(t) := \varphi^\infty(t) + \tfrac{2t-1}{1-t} \chi_{(\frac{1}{2},1)}(t).$$

We observe that $\varphi^p \in \Phi_s$ if and only if $p \geqslant 1$. Furthermore, $\varphi_{\max}, \varphi_{\exp}, \varphi^{\infty,2} \in \Phi_s$, $\varphi^\infty \in \Phi_c \setminus \Phi_s$ and $\varphi_{\sin} \in \Phi_w \setminus \Phi_c$.

Based on these examples, we can make a number of observations, which we record here.

– We observe that $\varphi^1 \simeq \varphi_{\sin}$ and $\varphi^\infty \simeq \varphi^{\infty,2}$. Therefore, neither Φ_c nor Φ_s is invariant under equivalence of Φ-functions. For the equivalence $\simeq$ see Definition 2.1.7.
– We observe that $\varphi^p \to \varphi^\infty$ and $\tfrac{1}{p}\varphi^{\infty,2} \to \varphi^\infty$ point-wise as $p \to \infty$. Therefore, Φ_s is not invariant under point-wise limits of Φ-functions.
– We note that $\min\{\varphi^1, \varphi^2\} \in \Phi_w \setminus \Phi_c$, so Φ_c is not preserved under point-wise minimum.

Fig. 2.2 Functions ψ (dashed) and $\psi^{\frac{1}{p}}$ (solid) with $p = 1.5$

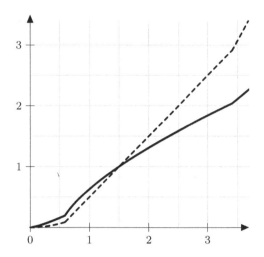

– Let $\psi(t) := \frac{1}{2} \max\{\varphi^2(t), 2t - 1\} = \max\{\frac{1}{4}t^2, t - \frac{1}{2}\}$. Then ψ is convex, but $\psi^{\frac{1}{p}}$ is not convex for any $p > 1$, see Fig. 2.2. Therefore, the improved convexity is lost by an inconsequential change. In contrast to this, both φ^2 and ψ satisfy (aInc)$_2$, so higher order ($p = 2$) "almost convexity" is preserved.

In a small number of results, we need left-continuity even though Φ_w is otherwise sufficient. The next result is one of them.

Lemma 2.1.5 *If $\varphi \in \Phi_w$ is left-continuous, then it is lower semicontinuous, i.e.*

$$\varphi(t) \leqslant \liminf_{t_k \to t} \varphi(t_k).$$

Proof Let $t_k \to t$ and $t'_k := \min\{t, t_k\}$. Then $t'_k \to t^-$ and since φ is left-continuous and increasing, we conclude that $\varphi(t) = \lim_{t_k \to t} \varphi(t'_k) \leqslant \liminf_{t_k \to t} \varphi(t_k)$. □

Remark 2.1.6 Let φ be a Φ-prefunction. As an increasing function, φ satisfies

$$\varphi\big(\inf A\big) \leqslant \inf_{t \in A} \varphi(t)$$

for every non-empty set $A \subset [0, \infty)$. If φ is not continuous, the inequality sign cannot in general be replaced by an equal-sign: with φ^∞ from Example 2.1.4 and $A := (1, \infty)$ we have $0 = \varphi^\infty(1) < \inf_{t \in A} \varphi^\infty(t) = \infty$.

However, for every $\lambda > 1$ we have

$$\inf \varphi(A) \leqslant \varphi\big(\lambda \inf A\big).$$

Indeed, if $\inf A = 0$, then the claim follows by $\lim_{t \to 0^+} \varphi(t) = 0$. If $\inf A > 0$, then $\inf A < \lambda \inf A$ so there exists $t \in A$ with $t < \lambda \inf A$. Now, the monotonicity of φ implies $\inf \varphi(A) \leqslant \varphi(\lambda \inf A)$.

Let φ satisfy (aInc)$_1$ with constant a. Let us show that

- $\varphi(\theta t) \leqslant a\theta\varphi(t)$ for all $\theta \in [0, 1]$ and $t \geqslant 0$; and
- $\varphi(\Theta t) \geqslant \frac{\Theta}{a}\varphi(t)$ for all $\Theta \in [1, \infty)$ and $t \geqslant 0$.

This is clear for $t = 0$, so we assume that $t > 0$. Since φ satisfies (aInc)$_1$ and $\theta t \leqslant t \leqslant \Theta t$, we obtain that

$$\frac{1}{a}\frac{\varphi(\theta t)}{\theta t} \leqslant \frac{\varphi(t)}{t} \leqslant a\frac{\varphi(\Theta t)}{\Theta t}.$$

Both claims follow from these inequalities.

Definition 2.1.7 Two functions φ and ψ are *equivalent*, $\varphi \simeq \psi$, if there exists $L \geqslant 1$ such that $\varphi(\frac{t}{L}) \leqslant \psi(t) \leqslant \varphi(Lt)$ for all $t \geqslant 0$.

This is the correct notion of equivalence for Φ-functions, as can be seen from Remarks 2.2.4 and 2.4.4. Later in Proposition 3.2.4 we show that equivalent Φ-functions give the same space with comparable (quasi)norms.

We observe that $\simeq$ is an equivalence relation in a set of functions from $[0, \infty)$ to $[0, \infty]$:

- $\varphi \simeq \varphi$ for every φ (reflexivity);
- $\varphi \simeq \psi$ implies $\psi \simeq \varphi$ (symmetry);
- $\varphi_1 \simeq \varphi_2$ and $\varphi_2 \simeq \varphi_3$ imply $\varphi_1 \simeq \varphi_3$ (transitivity).

The following lemma and example illustrate the reason for using (aInc) and (aDec) rather than (Inc) and (Dec): the former are compatible with equivalence of Φ-functions, whereas the latter are not. (In addition, the "almost" part often make the assumptions easier to check, e.g. in Proposition 3.6.2.) For future reference we also record the fact that the values of a Φ-prefunction at 0 and ∞ are preserved by equivalence.

Lemma 2.1.8 *Let* $\varphi, \psi : [0, \infty) \to [0, \infty]$ *be increasing with* $\varphi \simeq \psi$.

(a) *If φ is a Φ-prefunction, then ψ is a Φ-prefunction.*
(b) *If φ satisfies* (aInc)$_p$, *then ψ satisfies* (aInc)$_p$.
(c) *If φ satisfies* (aDec)$_q$, *then ψ satisfies* (aDec)$_q$.

Proof For (a), we assume that φ is a Φ-prefunction. Let $L \geqslant 1$ be such that $\varphi(\frac{t}{L}) \leqslant \psi(t) \leqslant \varphi(Lt)$. With $t = 0$, this gives $\psi(0) = 0$. Furthermore,

$$0 \leqslant \lim_{t \to 0^+} \psi(t) \leqslant \lim_{t \to 0^+} \varphi(Lt) = \lim_{t \to 0^+} \varphi(t) = 0$$

and

$$\lim_{t\to\infty} \psi(t) \geqslant \lim_{t\to\infty} \varphi(\tfrac{t}{L}) = \lim_{t\to\infty} \varphi(t) = \infty.$$

We show only (b), as the proof of (c) is similar but simpler. Let $0 < s < t$ and assume first that $L^2 s < t$, where L is the constant from the equivalence. By (aInc)$_p$ of φ with constant a, we obtain

$$\frac{\psi(s)}{s^p} \leqslant \frac{\varphi(Ls)}{s^p} = L^p \frac{\varphi(Ls)}{(Ls)^p} \leqslant aL^p \frac{\varphi(t/L)}{(t/L)^p} \leqslant aL^{2p} \frac{\psi(t)}{t^p}.$$

Assume then that $t \in (s, L^2 s]$. Using that ψ is increasing, we find that

$$\frac{\psi(s)}{s^p} \leqslant \frac{\psi(t)}{s^p} = \frac{t^p}{s^p} \frac{\psi(t)}{t^p} \leqslant L^{2p} \frac{\psi(t)}{t^p}.$$

We have shown that ψ satisfies (aInc)$_p$ with constant aL^{2p}.   □

The next example shows that (Inc)$_p$ without the "almost" is not invariant under equivalence of Φ-functions.

Example 2.1.9 Let $\varphi(t) := t^2$ and $\psi(t) := t^2 + \max\{t-1, 0\}$ for $t \geqslant 0$, see Fig. 2.3. Then $\varphi(t) \leqslant \psi(t) \leqslant \varphi(2t)$ so that $\varphi \simeq \psi$. Clearly φ satisfies (Inc)$_2$. Suppose that ψ satisfies (Inc)$_p$ for $p \geqslant 1$. We write the condition at points $t = 2$ and $s = 3$:

$$\frac{4+1}{2^p} = \frac{\psi(2)}{2^p} \leqslant \frac{\psi(3)}{3^p} = \frac{9+2}{3^p},$$

i.e. $(\tfrac{3}{2})^p \leqslant \tfrac{11}{5}$. This means that $p < 2$, hence ψ does not satisfy (Inc)$_2$.

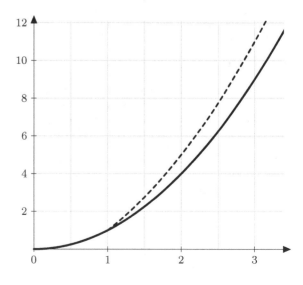

Fig. 2.3 Functions φ (solid) and ψ (dashed) from Example 2.1.9

Recall, that $f \approx g$ means that $c_1 f(x) \leqslant g(x) \leqslant c_2 f(x)$ for all relevant values of x. Next we consider the two notions of equivalence, $\simeq$ and $\approx$. The difference is illustrated by the fact that $\varphi^\infty \simeq \varphi^{\infty,2}$ but $\varphi^\infty \not\approx \varphi^{\infty,2}$ from our earlier examples.

Lemma 2.1.10 *Let $\varphi, \psi : [0, \infty) \to [0, \infty]$.*

(a) *If φ satisfies (aInc)$_1$ and $\varphi \approx \psi$, then $\varphi \simeq \psi$ and ψ satisfies (aInc)$_1$.*
(b) *If φ satisfies (aDec), then $\varphi \simeq \psi$ implies $\varphi \approx \psi$.*

Proof Assume first that $\varphi \approx \psi$ with constant $L \geqslant 1$. Then $\psi(t) \leqslant L\varphi(t) \leqslant \varphi(aLt)$ by (aInc)$_1$. The lower bound is similar, and thus $\varphi(t) \simeq \psi(t)$. For (aInc)$_1$ we argue as follows:

$$\frac{\psi(s)}{s} \leqslant L\frac{\varphi(s)}{s} \leqslant aL\frac{\varphi(t)}{t} \leqslant aL^2\frac{\psi(t)}{t}$$

for $0 < s < t$. Note that here we do not need the function to be increasing, in contrast to Lemma 2.1.8.

Assume next that $\varphi \simeq \psi$ and φ satisfies (aDec)$_q$. Then $\psi(t) \leqslant \varphi(Lt) \leqslant aL^q\varphi(t)$ by (aDec)$_q$. The lower bound is similar, and thus $\varphi(t) \approx \psi(t)$.   □

2.2 Upgrading Φ-Functions

We have seen in the previous section that weak Φ-functions have much better invariance properties than convex or strong Φ-functions. However, in many cases it is nicer to work with Φ-functions of the latter classes. This can often be achieved by upgrading the Φ-function that we obtain from some process. The tools for doing so are developed in this section. An alternative approach to upgrading is to use the conjugate function; here we present a direct method, whereas the conjugate function is used in Sect. 2.5.

Note that we can always estimate a weak Φ-function as follows:

$$\varphi(s + t) \leqslant \varphi(2 \max\{s, t\}) \leqslant \varphi(2s) + \varphi(2t).$$

Next we consider more sophisticated convexity-type properties.

Lemma 2.2.1 *If $\varphi \in \Phi_w$ satisfies (aInc)$_p$ with $p \geqslant 1$, then there exists $\psi \in \Phi_c$ equivalent to φ such that $\psi^{1/p}$ is convex. In particular, ψ satisfies (Inc)$_p$.*

Proof Assume first that φ satisfies (aInc)$_1$. Let us define $\varphi_2(0) := 0$ and

$$\varphi_2(t) := t \sup_{s \in (0,t]} \frac{\varphi(s)}{s}$$

for $t > 0$. Then φ_2 is increasing and satisfies (Inc)$_1$. With the choice $s = t > 0$ in the supremum, we obtain $\varphi \leqslant \varphi_2$. If $s \in (0, t]$, then by (aInc)$_1$ we have $\frac{\varphi(s)}{s} \leqslant a\frac{\varphi(t)}{t}$, so that

$$\varphi_2(t) = t \sup_{s \in [0,t]} \frac{\varphi(s)}{s} \leqslant ta \frac{\varphi(t)}{t} = a\varphi(t).$$

Moreover, $\varphi(0) = 0 = \varphi_2(0)$. Thus it follows from Lemma 2.1.10(a) that $\varphi \simeq \varphi_2$.
Note that $\frac{\varphi_2(s)}{s}$ is measurable since it is increasing. Let us define

$$\psi(t) := \int_0^t \frac{\varphi_2(s)}{s} ds.$$

Since ψ is defined by an integral of a non-negative function, it is continuous in a
closure of $\{\psi < \infty\}$. Moreover, ψ is continuous in the set $\{\psi = \infty\}$. Thus ψ is
left-continuous in $[0, \infty)$.

We define $s_\infty := \inf\{s : \varphi_2(s) = \infty\}$. In $(0, s_\infty)$, we have that $\psi'(t) = \frac{\varphi_2(t)}{t}$ is
increasing. Since ψ is continuous this yields that ψ is convex in $[0, s_\infty]$. In (s_∞, ∞)
we have $\psi = \infty$. These together imply that ψ is convex in $[0, \infty)$. Since φ_2 satisfies
(Inc)$_1$ we obtain that

$$\varphi_2(\tfrac{t}{2}) = \int_{t/2}^t \frac{\varphi_2(t/2)}{t/2} ds \leqslant \int_{t/2}^t \frac{\varphi_2(s)}{s} ds \leqslant \psi(t) \leqslant \varphi_2(t)$$

and hence $\varphi \simeq \varphi_2 \simeq \psi$. Thus ψ is a Φ-prefunction, by Lemma 2.1.8(a). This
concludes the proof of $\psi \in \Phi_c$ in the case $p = 1$.

Assume next that φ satisfies (aInc)$_p$ with $p > 1$. Then $\varphi(t)^{1/p}$ satisfies (aInc)$_1$.
Hence by the first part of the proof there exists $\xi \in \Phi_c$ such that $\xi \simeq \varphi^{1/p}$. Set
$\psi := \xi^p$. A convex function to a power greater than one (i.e. ξ^p) is convex, so that
$\psi \in \Phi_c$ and further $\psi \simeq \varphi$, as required. Let $p \in [1, \infty)$. Since $\psi^{1/p}$ is convex,
$\frac{\psi(t)^{1/p}}{t}$ is increasing by (2.1.1). Hence $\frac{\psi(t)}{t^p}$ is increasing, and so (Inc)$_p$ holds. □

Next we show that weak Φ-functions still have a quasi-convexity property, even
though they are not convex.

Corollary 2.2.2 *If $\varphi \in \Phi_w$, then there exists a constant $\beta > 0$ such that*

$$\varphi\left(\beta \sum_{k=1}^\infty a_k w_k\right) \leqslant \sum_{k=1}^\infty \varphi(a_k) w_k$$

for all a_k, $w_k \geqslant 0$ with $\sum w_k = 1$.

Proof The result is well-known for convex functions, but we provide a proof in
the interest of completeness. So suppose first that $\psi \in \Phi_c$. Denote $w'_{m+1} :=$

$\sum_{k=m+1}^{\infty} w_w$ and $a'_{m+1} = 0$. Then by convexity

$$\psi\left(\sum_{k=1}^{m} a_k w_k\right) = \psi\left(\sum_{k=1}^{m} a_k w_k + a'_{m+1} w'_{m+1}\right)$$

$$\leqslant \sum_{k=1}^{m} \psi(a_k) w_k + \psi(a'_{m+1}) w'_{m+1} \leqslant \sum_{k=1}^{\infty} \psi(a_k) w_k.$$

The inequality follows with $\beta = 1$ by left-continuity as $m \to \infty$.

Let then $\varphi \in \Phi_w$. By Lemma 2.2.1, there exists $\psi \in \Phi_c$ such that $\varphi \simeq \psi$ with constant $L \geqslant 1$. Choose $\beta := L^{-2}$. Then

$$\varphi\left(\beta \sum_{k=1}^{\infty} a_k w_k\right) \leqslant \psi\left(\frac{1}{L} \sum_{k=1}^{\infty} a_k w_k\right) \leqslant \sum_{k=1}^{\infty} \psi\left(\tfrac{1}{L} a_k\right) w_k \leqslant \sum_{k=1}^{\infty} \varphi(a_k) w_k. \qquad \square$$

We next show how to upgrade a weak Φ-function to a strong Φ-function. Most of the work was done in Lemma 2.2.1; it remains to show that an equivalent Φ-function is continuous into $[0, \infty]$. The next proof is based on the idea that changing a Φ-function in a compact subinterval of $(0, \infty)$ does not affect its equivalence class.

Theorem 2.2.3 *Every weak Φ-function is equivalent to a strong Φ-function.*

Proof Let φ be a weak Φ-function. Since φ satisfies (aInc)$_1$, Lemma 2.2.1 implies that there exists $\varphi_c \in \Phi_c$ with $\varphi_c \simeq \varphi$. Convexity implies that φ_c is continuous in the set $\{\varphi_c < \infty\}$. If this set equals $[0, \infty)$ we are done. Otherwise, denote $t_\infty := \inf\{t : \varphi_c(t) = \infty\} \in (0, \infty)$ and define

$$\varphi_s(t) := \varphi_c(t) + \frac{2t - t_\infty}{t_\infty - t} \chi_{(\frac{1}{2} t_\infty, t_\infty)}(t) + \infty \chi_{[t_\infty, \infty)}(t).$$

As the sum of three convex functions, φ_s is convex. Furthermore, $\varphi_s = \varphi_c$ in $[0, \frac{1}{2} t_\infty] \cup [t_\infty, \infty)$. Hence $\varphi_s(\frac{t}{2}) \leqslant \varphi_c(t) \leqslant \varphi_s(t)$, so that $\varphi_s \simeq \varphi_c \simeq \varphi$. Since φ_s is increasing we obtain by Lemma 2.1.8(a) that φ_s is a Φ-prefunction. Since also $\lim_{t \to t_\infty} \varphi_s(t) = \infty$, we conclude that φ_s is the required strong Φ-function. $\qquad \square$

Remark 2.2.4 The functions φ^∞ and $\varphi^{\infty,2}$ from Example 2.1.4 show that the previous theorem is not true with the $\approx$-equivalence. Indeed, if $\psi \approx \varphi^\infty$, then necessarily $\psi = \varphi^\infty$.

Definition 2.2.5 We say that a function $\varphi : [0, \infty) \to [0, \infty]$ satisfies Δ_2, or is *doubling* if there exists a constant $K \geqslant 2$ such that

$$\varphi(2t) \leqslant K\varphi(t) \quad \text{for all} \quad t \geqslant 0.$$

Next we show that (aDec) is a quantitative version of doubling.

Lemma 2.2.6

(a) *If $\varphi \in \Phi_w$, then Δ_2 is equivalent to* (aDec).
(b) *If $\varphi \in \Phi_c$, then Δ_2 is equivalent to* (Dec).

Proof Assume (aDec) holds. Then there exists $q > 1$ such that

$$\frac{\varphi(2t)}{(2t)^q} \leqslant a \frac{\varphi(t)}{t^q},$$

so $\varphi(2t) \leqslant a2^q\varphi(t)$. Thus (aDec) implies Δ_2 with $K = 2^q a$.

Assume then that φ satisfies Δ_2 and let $0 < s < t$. Choose an integer $k \geqslant 1$ such that $2^{k-1}s < t \leqslant 2^k s$. Then

$$\varphi(t) \leqslant \varphi(2^k s) \leqslant K\varphi(2^{k-1}s) \leqslant \cdots \leqslant K^k\varphi(s).$$

We define $q := \log_2(K) \geqslant 1$. Then the previous inequality and $t > 2^{k-1}s$ yield that

$$\frac{\varphi(t)}{t^q} \leqslant K^k \frac{\varphi(s)}{t^q} \leqslant K^k \frac{\varphi(s)}{2^{q(k-1)}s^q} = K \frac{\varphi(s)}{s^q}.$$

Thus φ satisfies (aDec)$_q$. Hence (a) is proved.

For (b) we are left to show that convexity and Δ_2 yield (Dec) for some $q \geqslant 1$. Let $t \geqslant 2s$ and q be the exponent defined in (a). By case (a),

$$\frac{\varphi(t)}{\varphi(s)} \leqslant K \left(\frac{t}{s}\right)^q \leqslant \left(\frac{t}{s}\right)^{2q}$$

and (Dec) holds for $t \geqslant 2s$ with any exponent that is at least $2q$. Suppose next that $t \in (s, 2s)$. Choose $\theta := \frac{t}{s} - 1 \in (0, 1)$ and note that $t = (1 - \theta)s + \theta 2s$. By convexity and Δ_2, we find that

$$\varphi(t) \leqslant (1 - \theta)\varphi(s) + \theta\varphi(2s) \leqslant (1 - \theta + K\theta)\varphi(s).$$

Therefore by the generalized Bernoulli inequality in the second step we obtain

$$\frac{\varphi(t)}{\varphi(s)} \leqslant 1 + (K - 1)\theta \leqslant (1 + \theta)^{K-1} = \left(\frac{t}{s}\right)^{K-1}.$$

Combining the two cases, we see that (Dec)$_{q_2}$ holds with $q_2 := \max\{2q, K - 1\}$. □

By the previous lemma it follows that if $\varphi \in \Phi_c$ satisfies (aDec)$_q$, then it satisfies (Dec)$_{q_2}$ for some possibly larger q_2. The previous lemma and Lemma 2.1.8(c) yield that if φ satisfies Δ_2 and $\varphi \simeq \psi$, then ψ satisfies Δ_2.

Proposition 2.2.7 *If $\varphi \in \Phi_w$ satisfies* (aDec), *then there exists $\psi \in \Phi_s$ with $\psi \approx \varphi$ which is a strictly increasing bijection.*

Proof By Lemma 2.2.1, there exists $\psi \in \Phi_c$ with $\psi \simeq \varphi$. Then ψ satisfies (aDec) by Lemma 2.1.8(c) and $\psi \approx \varphi$ by Lemma 2.1.10. If $\psi(t) = 0$ for some $t > 0$, then it follows from (aDec) that $\psi \equiv 0$, which contradicts $\psi \in \Phi_w$. The same holds if $\psi(t) = \infty$ for some t. Hence $\psi(t) \in (0, \infty)$ for every $t > 0$. Then convexity implies that ψ is continuous and strictly increasing.   □

2.3 Inverse Φ-Functions

Since our weak Φ-functions are not bijections, they are not strictly speaking invertible. However, we can define a left-continuous function with many properties of the inverse, which we call for simplicity left-inverse. Note that this is not the left-inverse in the sense of abstract algebra. For the elegance of our presentation, we extend in this section all Φ-functions to the interval $[0, \infty]$ by $\varphi(\infty) := \infty$.

Definition 2.3.1 By $\varphi^{-1} : [0, \infty] \rightarrow [0, \infty]$ we denote the *left-inverse* of $\varphi : [0, \infty] \rightarrow [0, \infty]$,

$$\varphi^{-1}(\tau) := \inf\{t \geqslant 0 : \varphi(t) \geqslant \tau\}.$$

In the Lebesgue case, the inverse of $t \mapsto t^p$ is $t \mapsto t^{\frac{1}{p}}$. As a more general intuition, this means that we flip (mirror) the function over the line $y = x$, and choose the value of the discontinuities so as to make the function left-continuous. This is illustrated in the following example.

Example 2.3.2 We define $\varphi : [0, \infty] \rightarrow [0, \infty]$ by

$$\varphi(t) := \begin{cases} 0 & \text{if } t \in [0, 2] \\ t - 2 & \text{if } t \in (2, 4] \\ 3 & \text{if } t \in (4, 6] \\ t - 3 & \text{if } t \in (6, \infty], \end{cases}$$

see Fig. 2.4. Then $\varphi \in \Phi_w \setminus \Phi_c$ and the left-inverse is given by

$$\varphi^{-1}(t) = \begin{cases} 0 & \text{if } t = 0 \\ t + 2 & \text{if } t \in (0, 2] \\ 4 & \text{if } t \in (2, 3] \\ t + 3 & \text{if } t \in (3, \infty]. \end{cases}$$

Fig. 2.4 A weak Φ-function
(solid) and its left-inverse
(dashed)

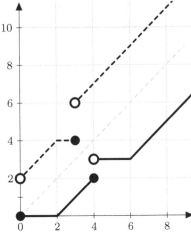

Fig. 2.5 t_0 and t_∞

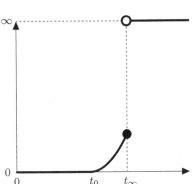

With these expressions we can calculate the compositions $\varphi \circ \varphi^{-1}$ and $\varphi^{-1} \circ \varphi$:

$$\varphi(\varphi^{-1}(t)) := \begin{cases} t & \text{if } t \in [0,2] \\ 2 & \text{if } t \in (2,3] \\ t & \text{if } t \in (3,\infty], \end{cases} \qquad \varphi^{-1}(\varphi(t)) := \begin{cases} 0 & \text{if } t \in [0,2] \\ t & \text{if } t \in (2,4] \\ 4 & \text{if } t \in (4,6] \\ t & \text{if } t \in (6,\infty]. \end{cases}$$

We next investigate when the composition of φ and φ^{-1} is the identity. Note that the following result holds only for convex Φ-functions. In the set $\Phi_w \setminus \Phi_c$ the behaviour of the composition is more complicated, as indicated by the previous example. In the proof we use $t_0 := \sup\{t : \varphi(t) = 0\}$ and $t_\infty := \inf\{t : \varphi(t) = \infty\}$ which are illustrated in Fig. 2.5.

Lemma 2.3.3 *Let* $\varphi \in \Phi_c$, $t_0 := \sup\{t : \varphi(t) = 0\}$ *and* $t_\infty := \inf\{t : \varphi(t) = \infty\}$. *Then*

$$
\varphi^{-1}(\varphi(t)) = \begin{cases} 0, & t \leqslant t_0, \\ t, & t_0 < t \leqslant t_\infty, \\ t_\infty & t > t_\infty. \end{cases} \quad and \quad \varphi(\varphi^{-1}(\tau)) = \min\{\tau, \varphi(t_\infty)\}.
$$

In particular, if $\varphi \in \Phi_s$, *then* $\varphi(\varphi^{-1}(s)) \equiv s$.

Note that the last property means that φ^{-1} is in fact the right-inverse of φ in the sense of abstract algebra when $\varphi \in \Phi_s$.

Proof We start with the composition $\varphi^{-1} \circ \varphi$. Since φ is left-continuous, $\varphi(t_0) = 0$. Hence $\varphi(t) = 0$ for all $0 \leqslant t \leqslant t_0$ and thus $\varphi^{-1}(\varphi(t)) = 0$ for $t \leqslant t_0$. Assume then that $t_0 < t \leqslant t_\infty$. Since φ is convex and positive on $(t_0, t_\infty]$, it is strictly increasing there, so that φ^{-1} is the usual inverse and hence $\varphi^{-1}(\varphi(t)) = t$. Assume finally that $t > t_\infty$. Then $\varphi(t) = \infty$ and thus $\varphi^{-1}(\varphi(t)) = t_\infty$.

Then we consider the composition $\varphi \circ \varphi^{-1}$. If $\tau = 0$, then $\varphi(\varphi^{-1}(\tau)) = \varphi(0) = 0$, so the claim holds. If $\tau \in (0, \varphi(t_\infty)]$, then $\varphi^{-1}(\tau) \in (t_0, t_\infty]$. Since φ is strictly increasing and continuous in this interval, $\varphi(\varphi^{-1}(\tau)) = \tau$. Let then $\tau > \varphi(t_\infty)$. Since $\varphi(t) = \infty \geqslant \tau$ for all $t > t_\infty$, it follows from the definition that $\varphi^{-1}(\tau) \leqslant t_\infty$. On the other hand for $t < t_\infty$ we have $\varphi(t) \leqslant \varphi(t_\infty) < \tau$. These yield that $\varphi^{-1}(\tau) = t_\infty$ and so $\varphi(\varphi^{-1}(\tau)) = \varphi(t_\infty)$.

The claim for strong Φ-functions follows since $\varphi(t_\infty) = \infty$ in this case. □

In the previous proof we used the fact that every convex Φ-function is strictly increasing on $\varphi^{-1}(0, \infty) = (t_0, t_\infty)$. This also yields the following result.

Corollary 2.3.4 *Let* $\varphi \in \Phi_c$. *If* $\varphi(s) \in (0, \infty)$, *then* $\varphi^{-1}(\varphi(s)) = s$.

Of course, if $\varphi \in \Phi_c$ satisfies (aDec), then φ is bijective and φ^{-1} is just the regular inverse. Next we move to weak Φ-functions and prove the geometrically intuitive result that $^{-1}$ maps $\simeq$ to $\approx$.

Question 2.3.5 Does the next result hold for all Φ-prefunctions?

Theorem 2.3.6 *Let* $\varphi, \psi \in \Phi_w$. *Then* $\varphi \simeq \psi$ *if and only if* $\varphi^{-1} \approx \psi^{-1}$.

Proof Assume first that $\varphi \simeq \psi$, i.e. $\varphi(\frac{t}{L}) \leqslant \psi(t) \leqslant \varphi(Lt)$ for all $t \geqslant 0$. Then

$$
\psi^{-1}(\tau) = \inf\{t \geqslant 0 : \psi(t) \geqslant \tau\} \geqslant \inf\{t \geqslant 0 : \varphi(Lt) \geqslant \tau\} = \frac{1}{L}\varphi^{-1}(\tau)
$$

and

$$
\psi^{-1}(\tau) = \inf\{t \geqslant 0 : \psi(t) \geqslant \tau\} \leqslant \inf\{t \geqslant 0 : \varphi(\frac{t}{L}) \geqslant \tau\} = L\varphi^{-1}(\tau).
$$

Thus $\varphi \simeq \psi$ implies $\varphi^{-1} \approx \psi^{-1}$.

Assume then that $\varphi^{-1} \approx \psi^{-1}$. By Theorem 2.2.3 there exist $\varphi_s, \psi_s \in \Phi_s$ such that $\varphi_s \simeq \varphi$ and $\psi_s \simeq \psi$. By the first part of the proof, $\varphi_s^{-1} \approx \varphi^{-1}$ and $\psi_s^{-1} \approx \psi^{-1}$ so that $\varphi_s^{-1} \approx \psi_s^{-1}$ by transitivity of $\approx$. If we show that this implies $\varphi_s \simeq \psi_s$, then the claim follows, since "$\simeq$" is an equivalence relation.

Let $t_0 := \sup\{t \,:\, \varphi_s(t) = 0\}$ and $t_\infty := \inf\{t \,:\, \varphi_s(t) = \infty\}$. Let us first assume that $t \in (t_0, t_\infty)$. We obtain by Corollary 2.3.4 and $\varphi_s^{-1} \approx \psi_s^{-1}$ that

$$\tfrac{1}{L}t = \tfrac{1}{L}\varphi_s^{-1}(\varphi_s(t)) \leqslant \psi_s^{-1}(\varphi_s(t)) \leqslant L\varphi_s^{-1}(\varphi_s(t)) = Lt.$$

Then we take ψ_s of both sides and use Lemma 2.3.3 to obtain that $\psi_s(\tfrac{1}{L}t) \leqslant \psi_s(\psi_s^{-1}(\varphi_s(t))) = \varphi_s(t) \leqslant \psi_s(Lt)$. We have shown the claim for $t \in (t_0, t_\infty)$. By continuity, $\psi_s(\tfrac{1}{L}t_0) = \lim_{t \to t_0^+} \psi_s(\tfrac{1}{L}t) \leqslant \lim_{t \to t_0^+} \varphi_s(t) = \varphi_s(t_0) = 0$ and hence $\psi_s(\tfrac{1}{L}t) = 0$ for $t \in (0, t_0]$. The inequality $\varphi(t) \leqslant \psi(Lt)$ is clear, since $\varphi(t) = 0$ when $t \leqslant t_0$. Similarly, we prove that $\psi_s(\tfrac{1}{L}t) \leqslant \infty \leqslant \psi_s(Lt)$ when $t \geqslant t_\infty$. $\quad\square$

Proposition 2.3.7 *Let $\varphi \in \Phi_w$ and $p, q > 0$. Then*

(a) *φ satisfies (aInc)$_p$ if and only if φ^{-1} satisfies (aDec)$_{\frac{1}{p}}$.*

(b) *φ satisfies (aDec)$_q$ if and only if φ^{-1} satisfies (aInc)$_{\frac{1}{q}}$.*

Proof Suppose first that $\varphi \in \Phi_s$. Let $\tau \in (0, \infty)$ and $t_\infty := \inf\{t \,:\, \varphi(t) = \infty\}$. Since φ is surjection, there exists $t \in (0, t_\infty)$ such that $\varphi(t) = \tau$. We obtain by Corollary 2.3.4 that

$$\frac{\varphi^{-1}(\tau)}{\tau^{\frac{1}{p}}} = \frac{\varphi^{-1}(\varphi(t))}{\varphi(t)^{\frac{1}{p}}} = \left(\frac{\varphi(t)}{t^p}\right)^{-\frac{1}{p}}.$$

Hence the fraction on the left-hand side is almost decreasing if and only if the fraction on the right-hand side is almost increasing, and vice versa. This proves the claim for φ restricted to (t_0, t_∞). If $t \leqslant t_0$ or $t > t_\infty$, then $\varphi(t) = 0$ or $\varphi(t) = \infty$, and the claim is vacuous.

Consider then the general case $\varphi \in \Phi_w$. By Theorem 2.2.3 there exists $\psi \in \Phi_s$ with $\psi \simeq \varphi$. By Theorem 2.3.6, $\psi^{-1} \approx \varphi^{-1}$, so the claim follows from the first part of the proof, by Lemma 2.1.8. $\quad\square$

We next consider properties of φ^{-1} without assuming that $\varphi \in \Phi_w$. This will allow us to characterize the set $\{\varphi^{-1} \,:\, \varphi \in \Phi_w\}$ and those functions for which $(\varphi^{-1})^{-1} = \varphi$.

Lemma 2.3.8 *Let $\varphi, \psi \,:\, [0, \infty] \to [0, \infty]$ be increasing. Then the following implications hold:*

$$\begin{aligned}\varphi \leqslant \psi &\quad\Rightarrow\quad \psi^{-1} \leqslant \varphi^{-1}, \\ \varphi^{-1} < \psi^{-1} &\quad\Rightarrow\quad \psi < \varphi.\end{aligned}$$

Proof Suppose that $\varphi \leqslant \psi$ and $\tau \geqslant 0$. Then

$$\{t \geqslant 0 \,:\, \varphi(t) \geqslant \tau\} \subset \{t \geqslant 0 \,:\, \psi(t) \geqslant \tau\},$$

and so it follows from the definition that $\varphi^{-1}(\tau) \leqslant \psi^{-1}(\tau)$. The second implication is the contrapositive of the first one, so it is logically equivalent. □

The next lemma collects results regarding φ^{-1} for general φ. Example 2.3.10 concerns the sharpness of some of these results.

Lemma 2.3.9 *Let* $\varphi : [0, \infty] \to [0, \infty]$, $t, \tau \geqslant 0$ *and* $\varepsilon > 0$.

(a) *Then* φ^{-1} *is increasing,* $\varphi^{-1}(0) = 0$, $\varphi^{-1}(\varphi(t)) \leqslant t$ *and* $\varphi(\varphi^{-1}(\tau) - \varepsilon) < \tau$ *when* $\varphi^{-1}(\tau) \geqslant \varepsilon$.
(b) *If* φ *is left-continuous with* $\varphi(0) = 0$, *then* $\varphi(\varphi^{-1}(\tau)) \leqslant \tau$.
(c) *If* φ *is increasing, then* φ^{-1} *is left-continuous,* $t \leqslant \varphi^{-1}(\varphi(t) + \varepsilon)$ *and* $\tau \leqslant \varphi(\varphi^{-1}(\tau) + \varepsilon)$.
(d) *If* φ *satisfies* (aInc)$_\varepsilon$, *then* $\varphi^{-1}(\varphi(t)) \approx t$ *when* $\varphi(t) \in (0, \infty)$.
(e) *If* φ *with* $\lim_{t \to 0^+} \varphi(t) = 0$ *satisfies* (aDec), *then* $\varphi(\varphi^{-1}(\tau)) \approx \tau$.

Proof Let $0 \leqslant \tau_1 < \tau_2$. Then

$$\{t \geqslant 0 \,:\, \varphi(t) \geqslant \tau_2\} \subset \{t \geqslant 0 \,:\, \varphi(t) \geqslant \tau_1\}$$

and hence $\varphi^{-1}(\tau_1) \leqslant \varphi^{-1}(\tau_2)$ so that φ^{-1} is increasing. For $\tau := \varphi(t)$, the variable t belongs to the set $\{s \geqslant 0 \,:\, \varphi(s) \geqslant \tau\}$ so that $\varphi^{-1}(\varphi(t)) \leqslant t$. Since $\varphi^{-1}(\tau)$ is the infimum of the set $\{t \geqslant 0 \,:\, \varphi(t) \geqslant \tau\}$, it follows that $\varphi^{-1}(\tau) - \varepsilon$ is not a member of the set, hence $\varphi(\varphi^{-1}(\tau) - \varepsilon) < \tau$.

To prove (b), we use the left-continuity of φ and let $\varepsilon \to 0^+$ in $\varphi(\varphi^{-1}(\tau) - \varepsilon) < \tau$ if $\varphi^{-1}(\tau) > 0$. If $\varphi^{-1}(\tau) = 0$, then $\varphi(\varphi^{-1}(\tau)) = 0$, so the claim also holds.

We then consider (c). Since φ is increasing, $\varphi(s) \geqslant \varphi(t) + \varepsilon$ implies $s \geqslant t$, and so it follows from the definition that $\varphi^{-1}(\varphi(t) + \varepsilon) \geqslant t$. If $\varphi(s) < \tau$, then $\varphi^{-1}(\tau) \geqslant s$; applying this to $s := \varphi^{-1}(\tau) + \varepsilon$, we arrive at a contradiction, and so $\varphi(s) \geqslant \tau$. Finally we prove left-continuity. Let $\tau > 0$ and $\tau_k \to \tau^-$. By (a), φ^{-1} is increasing and thus $\varphi^{-1}(\tau_k) \leqslant \varphi^{-1}(\tau)$ for every k. This implies that $\limsup_{k \to \infty} \varphi^{-1}(\tau_k) \leqslant \varphi^{-1}(\tau)$. Since φ and φ^{-1} are increasing, we obtain by the latter inequality in (c) that

$$\varphi\left(\liminf_{k \to \infty} \varphi^{-1}(\tau_k) + \varepsilon\right) \geqslant \liminf_{k \to \infty} \varphi(\varphi^{-1}(\tau_k) + \varepsilon) \geqslant \liminf_{k \to \infty} \tau_k = \tau.$$

This yields by part (a) that $\liminf_{k \to \infty} \varphi^{-1}(\tau_k) + \varepsilon \geqslant \varphi^{-1}(\tau)$ for every $\varepsilon > 0$, and thus $\liminf_{k \to \infty} \varphi^{-1}(\tau_k) \geqslant \varphi^{-1}(\tau)$ as required.

For (d), suppose that $\varphi(s) \geqslant \varphi(t) \in (0, \infty)$ and $s < t$. It follows from (aInc)$_\varepsilon$ that $\varphi(s) \leqslant a(\frac{s}{t})^\varepsilon \varphi(t)$ so that $s \geqslant a^{-1/\varepsilon} t$. We conclude that

$$\varphi^{-1}(\varphi(t)) = \inf\{s \geqslant 0 \,:\, \varphi(s) \geqslant \varphi(t)\} \geqslant a^{-1/\varepsilon} t.$$

In view of $\varphi^{-1}(\varphi(t)) \leqslant t$ from (a), we have shown that $\varphi^{-1}(\varphi(t)) \approx t$.

It remains to prove (e). We consider the set $\{t : \varphi(t) \geqslant \tau\}$ and its infimum $\varphi^{-1}(\tau) > 0$. Then there exist (t_i) in the set and (s_i) in the complement with $t_i \to \varphi^{-1}(\tau)^+$ and $s_i \to \varphi^{-1}(\tau)^-$. It follows form $(aDec)_q$ that

$$\tau \leqslant \varphi(t_i) \leqslant a\Big(\frac{t_i}{\varphi^{-1}(\tau)}\Big)^q \varphi(\varphi^{-1}(\tau)) \quad \text{and} \quad \tau > \varphi(s_i) \geqslant \frac{1}{a}\Big(\frac{\varphi^{-1}(\tau)}{s_i}\Big)^q \varphi(\varphi^{-1}(\tau)).$$

When $i \to \infty$, we obtain $\varphi(\varphi^{-1}(\tau)) \approx \tau$. Suppose next that $\varphi^{-1}(\tau) = 0$. Then there exist $t_i \to 0^+$ with $\varphi(t_i) \geqslant \tau$. Since $\lim_{t \to 0} \varphi(t) = 0$, it follows that $\tau = 0$, and so $\varphi(\varphi^{-1}(\tau)) \approx \tau$ holds in this case also.   □

Example 2.3.10 The next example shows that the left-continuity is crucial in Lemma 2.3.9(b). If

$$\varphi(t) := \begin{cases} 2t & \text{if } t \in [0, 1) \\ t+2 & \text{if } t \in [1, \infty], \end{cases}$$

then

$$\varphi^{-1}(t) := \begin{cases} \frac{1}{2}t & \text{if } t \in [0, 2) \\ 1 & \text{if } t \in [2, 3) \\ t-2 & \text{if } t \in [3, \infty], \end{cases}$$

and thus $\varphi(\varphi^{-1}(2)) = \varphi(1) = 3$.

Let us also note that the inequalities in the previous lemma may be strict. Let $\varphi^\infty(t) := \infty \chi_{(1,\infty)}(t)$ as in Example 2.1.4. Then $(\varphi^\infty)^{-1} = \chi_{(0,\infty]}$ and thus

$$(\varphi^\infty)^{-1}(\varphi^\infty(\tfrac{1}{2})) = (\varphi^\infty)^{-1}(0) = 0 < \tfrac{1}{2}$$

and $\varphi^\infty((\varphi^\infty)^{-1}(1)) = \varphi^\infty(1) = 0 < 1$.

Lemma 2.3.11 *Let $\varphi : [0, \infty] \to [0, \infty]$. Then φ is increasing and left-continuous if and only if $(\varphi^{-1})^{-1} = \varphi$.*

Proof Assume first that φ is increasing and left-continuous. Let $0 < \varepsilon < t$. Lemma 2.3.9(a) yields that $\varphi^{-1}(\tau) \leqslant \varphi^{-1}(\varphi(t-\varepsilon)) \leqslant t-\varepsilon$ for $0 \leqslant \tau \leqslant \varphi(t-\varepsilon)$. Thus we obtain that

$$(\varphi^{-1})^{-1}(t) = \inf\{\tau \geqslant 0 : \varphi^{-1}(\tau) \geqslant t\} \geqslant \varphi(t-\varepsilon).$$

Since φ is left-continuous, this yields $(\varphi^{-1})^{-1}(t) \geqslant \varphi(t)$ as $\varepsilon \to 0^+$. If $t = 0$, the inequality also holds.

By Lemma 2.3.9(c), $\varphi^{-1}(\varphi(t)+\varepsilon) \geqslant t$ and hence $(\varphi^{-1})^{-1}(t) \leqslant \varphi(t)+\varepsilon$. Letting $\varepsilon \to 0^+$ we obtain $(\varphi^{-1})^{-1}(t) \leqslant \varphi(t)$. We have shown that $(\varphi^{-1})^{-1} = \varphi$.

Assume then conversely that $(\varphi^{-1})^{-1} = \varphi$. By Lemma 2.3.9(a), φ^{-1} is increasing. Hence Lemma 2.3.9(c) implies that $(\varphi^{-1})^{-1}$ is left-continuous. Furthermore, it is increasing by Lemma 2.3.9(a). Since $\varphi = (\varphi^{-1})^{-1}$, also φ is increasing and left-continuous. $\square$

Definition 2.3.12 We say that $\xi : [0, \infty] \to [0, \infty]$ belongs to Φ_w^{-1} if it is increasing, left-continuous, satisfies (aDec)$_1$, $\xi(t) = 0$ if and only if $t = 0$, and, $\xi(t) = \infty$ if and only if $t = \infty$.

Let us denote be Φ_{w+} the set of left-continuous weak Φ-functions. We next show that Φ_w^{-1} characterizes inverses of Φ_{w+}-functions and that $^{-1}$ is an involution in Φ_{w+}.

Proposition 2.3.13 *The transformation* $\varphi \mapsto \varphi^{-1}$ *is a bijection from* Φ_{w+} *to* Φ_w^{-1}:

(a) *If* $\varphi \in \Phi_{w+}$, *then* $\varphi^{-1} \in \Phi_w^{-1}$ *and* $(\varphi^{-1})^{-1} = \varphi$.
(b) *If* $\xi \in \Phi_w^{-1}$, *then* $\xi^{-1} \in \Phi_{w+}$ *and* $(\xi^{-1})^{-1} = \xi$.

Proof Both $(\varphi^{-1})^{-1} = \varphi$ and $(\xi^{-1})^{-1} = \xi$ follow from Lemma 2.3.11, since φ and ξ are assumed to be increasing and left-continuous.

Let us prove the first claim of (a). Proposition 2.3.7(a) yields that φ^{-1} satisfies (aDec)$_1$ and Lemma 2.3.9(a) and (c) yield that it is increasing and left-continuous. If $\varphi^{-1}(\tau) = 0$, then $\varphi(t) \geqslant \tau$ for every $t > 0$. Since $\lim_{t \to 0} \varphi(t) = 0$, it follows that $\tau = 0$. Similarly we conclude from $\lim_{t \to \infty} \varphi(t) = \infty$ that $\varphi^{-1}(\tau) = \infty$ implies that $\tau = \infty$. We have proved that $\varphi^{-1} \in \Phi_w^{-1}$.

We then consider the first claim of (b). Since ξ is increasing, it follows from Lemma 2.3.9(a) and (c) that ξ^{-1} is increasing and left-continuous. Since $\xi(0) = 0$, we obtain $\xi^{-1}(0) = 0$. Let $\varepsilon > 0$. By monotonicity and Lemma 2.3.9(a),

$$0 \leqslant \xi^{-1}(t) \leqslant \xi^{-1}(\xi(\varepsilon)) \leqslant \varepsilon$$

for $t \in [0, \xi(\varepsilon))$ (note that this interval is non-empty by assumption). Hence $\lim_{t \to 0^+} \xi^{-1}(t) = 0$. Let $m > 0$. By Lemma 2.3.9(c),

$$\xi^{-1}(t) \geqslant \xi^{-1}(\xi(m) + 1) \geqslant m$$

for all $t > \xi(m) + 1$ and hence $\lim_{t \to \infty} \xi^{-1}(t) = \infty$. Thus ξ^{-1} is a Φ-prefunction.

Let $\sigma > 0$ and $s \in (\frac{1}{2}\xi^{-1}(\sigma), \xi^{-1}(\sigma))$ (notice that the interval is non-empty by assumption). Then $\xi(s) < \sigma$ by definition of the inverse. Let a be from (aDec)$_1$ of ξ and suppose first that $\tau \geqslant a\sigma$. By (aDec)$_1$ it follows that $\xi(\frac{\tau}{a\sigma}s) \leqslant \frac{\tau}{\sigma}\xi(s) < \tau$. Hence by the definition of the inverse

$$\xi^{-1}(\tau) \geqslant \frac{\tau}{a\sigma}s \geqslant \frac{\tau}{2a\sigma}\xi^{-1}(\sigma).$$

If $\tau \in (\sigma, a\sigma)$, then since ξ^{-1} is increasing,

$$\xi^{-1}(\tau) \geqslant \xi^{-1}(\sigma) \geqslant \frac{\tau}{\sigma a}\xi^{-1}(\sigma).$$

We have shown that ξ^{-1} satisfies $(\mathrm{aInc})_1$. This together with the previous properties yields that $\xi^{-1} \in \Phi_{\mathrm{w}+}$. ☐

Note that we could not use Proposition 2.3.7 in the previous proof to derive $(\mathrm{aInc})_1$, since that proposition assumes that $\varphi \in \Phi_{\mathrm{w}}$, which we are proving.

2.4　Conjugate Φ-Functions

Definition 2.4.1 Let $\varphi : [0, \infty) \to [0, \infty]$. We denote by φ^* the *conjugate function* of φ which is defined, for $u \geqslant 0$, by

$$\varphi^*(u) := \sup_{t \geqslant 0} \big(tu - \varphi(t)\big).$$

In the Lebesgue case $t \mapsto \frac{1}{p}t^p$, the conjugate is given by $t \mapsto \frac{1}{p'}t^{p'}$, where p' is the Hölder conjugate exponent. By definition of φ^*,

$$tu \leqslant \varphi(t) + \varphi^*(u) \tag{2.4.1}$$

for every $t, u \geqslant 0$. This is called *Young's inequality*.

Lemma 2.4.2 *If $\varphi \in \Phi_{\mathrm{w}}$, then $\varphi^* \in \Phi_{\mathrm{c}}$.*

Note that φ^* is convex and left-continuous even if φ is not. The previous lemma does not extend to strong Φ-functions: if $\varphi(t) := t$, then $\varphi \in \Phi_{\mathrm{s}}$ but $\varphi^*(t) = \infty\chi_{(1,\infty)}(t)$ belongs to $\Phi_{\mathrm{c}} \setminus \Phi_{\mathrm{s}}$.

Proof At the origin we have $\varphi^*(0) = \sup_{t \geqslant 0}\big(-\varphi(t)\big) = \varphi(0) = 0$. If $u < v$, then

$$\varphi^*(u) = \sup_{t \geqslant 0}\big(tu - \varphi(t)\big) \leqslant \sup_{t \geqslant 0}\big(t(v-u) + tu - \varphi(t)\big) = \varphi^*(v)$$

and hence φ^* is increasing. It follows that $\varphi^*(t) \geqslant \varphi^*(0) = 0$ for all $t > 0$.

Since $\lim_{t \to 0}\varphi(t) = 0$, there exists $t_1 > 0$ with $\varphi(t_1) < \infty$. Then $\varphi^*(u) \geqslant ut_1 - \varphi(t_1)$ so that $\lim_{u \to \infty}\varphi^*(u) = \infty$. Likewise, there exists $t_2 > 0$ with $\varphi(t_2) > 0$. It follows from $(\mathrm{aInc})_1$ that $\varphi(t) \geqslant \frac{\varphi(t_2)}{at_2}t$ when $t \geqslant t_2$. Hence

$$\varphi^*(u) \leqslant \max\left\{t_2 u, \sup_{t > t_2}(tu - \frac{\varphi(t_2)t}{at_2})\right\}.$$

When $u < \frac{\varphi(t_2)}{at_2}$, the second term is negative, and we see that $\lim_{u\to 0^+} \varphi^*(u) = 0$. Hence φ^* is a Φ-prefunction.

Next we show that φ^* is convex. For that let $\theta \in (0, 1)$ and $u, v \geqslant 0$. We obtain

$$\varphi^*(\theta u + (1 - \theta)v) = \sup_{t \geqslant 0} \left(t(\theta u + (1 - \theta)v) - \varphi(t)\right)$$

$$= \sup_{t \geqslant 0} \left(\theta(tu - \varphi(t)) + (1 - \theta)(tv - \varphi(t))\right)$$

$$\leqslant \theta \sup_{t \geqslant 0} \left(tu - \varphi(t)\right) + (1 - \theta) \sup_{t \geqslant 0} \left(tv - \varphi(t)\right)$$

$$= \theta \varphi^*(u) + (1 - \theta)\varphi^*(v).$$

Let $u_\infty^* := \inf\{u > 0 : \varphi^*(u) = \infty\}$. Convexity implies continuity in $[0, u_\infty^*)$ and continuity is clear in (u_∞^*, ∞). So we are left to show left-continuity at u_∞^*. For every $\lambda \in (0, 1)$ we have $\varphi^*(\lambda u_\infty^*) \leqslant \varphi^*(u_\infty^*)$ and hence $\limsup_{\lambda \to 1^-} \varphi^*(\lambda u_\infty^*) \leqslant \varphi^*(u_\infty^*)$. Let $m < \varphi^*(u_\infty^*)$ and t_1 be such that $t_1 u_\infty^* - \varphi(t_1) \geqslant m$. Then

$$\liminf_{\lambda \to 1^-} \varphi^*(\lambda u_\infty^*) \geqslant \liminf_{\lambda \to 1^-} \left(t_1 \lambda u_\infty^* - \varphi(t_1)\right) = t_1 u_\infty^* - \varphi(t_1) \geqslant m.$$

When $m \to \varphi^*(u_\infty^*)^-$ we obtain $\liminf_{\lambda \to 1^-} \varphi^*(\lambda u_\infty^*) \geqslant \varphi^*(u_\infty^*)$ and thus φ^* is left-continuous. □

Lemma 2.4.3 *Let* $\varphi, \psi : [0, \infty) \to [0, \infty]$ *and* $a, b > 0$.

(a) *If* $\varphi \leqslant \psi$, *then* $\psi^* \leqslant \varphi^*$.
(b) *If* $\psi(t) = a\varphi(bt)$ *for all* $t \geqslant 0$, *then* $\psi^*(u) = a\varphi^*(\frac{u}{ab})$ *for all* $u \geqslant 0$.
(c) *If* $\varphi \simeq \psi$, *then* $\varphi^* \simeq \psi^*$.

Proof We begin with the proof of (a). Let $\varphi(t) \leqslant \psi(t)$ for all $t \geqslant 0$. Then

$$\psi^*(u) = \sup_{t \geqslant 0} \left(tu - \psi(t)\right) \leqslant \sup_{t \geqslant 0} \left(tu - \varphi(t)\right) = \varphi^*(u)$$

for all $u \geqslant 0$. For the proof of (b), let $a, b > 0$ and $\psi(t) = a\varphi(bt)$ for all $t \geqslant 0$. Then

$$\psi^*(u) = \sup_{t \geqslant 0} \left(tu - \psi(t)\right) = \sup_{t \geqslant 0} \left(tu - a\varphi(bt)\right)$$

$$= \sup_{t \geqslant 0} a\left(bt\frac{u}{ab} - \varphi(bt)\right) = a\varphi^*\left(\frac{u}{ab}\right)$$

for all $u \geqslant 0$. For (c) assume that $\varphi(t/L) \leqslant \psi(t) \leqslant \varphi(Lt)$. By (b) and (a), we obtain that

$$\varphi^*(Lt) = (\varphi(\tfrac{t}{L}))^* \geqslant \psi^*(t) \geqslant (\varphi(Lt))^* = \varphi^*(\tfrac{t}{L}). \qquad \Box$$

Fig. 2.6 Sketch of the case
$\varphi^{**}(t_0) < \varphi(t_0)$

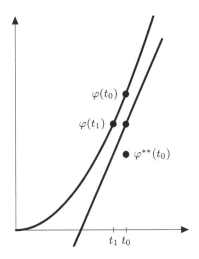

Remark 2.4.4 In (c) of the previous lemma, it is crucial that we have $\simeq$-equivalence, since the claim is false for $\approx$. Consider for instance $\varphi(t) = t$ and $\psi(t) = 2t$. Then clearly $\varphi \approx \psi$ and $\varphi \simeq \psi$. However, $\varphi^* = \infty\chi_{(1,\infty)}$ and $\psi^* = \infty\chi_{(2,\infty)}$, so that $\varphi^* \not\approx \psi^*$.

We denote $\varphi^{**} = (\varphi^*)^*$. For $\varphi \in \Phi_{\mathrm{w}} \setminus \Phi_{\mathrm{c}}$ we have $\varphi^{**} \neq \varphi$, since φ^{**} is convex and left-continuous (Lemma 2.4.2) but φ is not. However, on Φ_{c}, the operation * is an involution.

Proposition 2.4.5 *Let* $\varphi \in \Phi_{\mathrm{w}}$. *Then* $\varphi^{**} \simeq \varphi$ *and* φ^{**} *is the greatest convex minorant of* φ.

In particular, if $\varphi \in \Phi_{\mathrm{c}}$, *then* $\varphi^{**} = \varphi$ *and*

$$\varphi(t) = \sup_{u \geqslant 0} \left(tu - \varphi^*(u) \right) \quad \text{for all} \quad t \geqslant 0.$$

Proof Let us first assume that $\varphi \in \Phi_{\mathrm{c}}$ and prove the latter part of the proposition. By Lemma 2.4.2 we have $\varphi^{**} \in \Phi_{\mathrm{c}}$. By definition of φ^{**} and Young's inequality (2.4.1) we obtain

$$\varphi^{**}(t) = \sup_{u \geqslant 0} \left(ut - \varphi^*(u) \right) \leqslant \sup_{u \geqslant 0} \left(\varphi(t) + \varphi^*(u) - \varphi^*(u) \right) = \varphi(t). \qquad (2.4.2)$$

It remains to show $\varphi^{**}(t) \geqslant \varphi(t)$. We prove this by contradiction. Assume to the contrary that there exists $t_0 \geqslant 0$ with $\varphi^{**}(t_0) < \varphi(t_0)$, see Fig. 2.6. Suppose first that $\varphi(t_0) < \infty$. Since φ is left-continuous, there exists $t_1 < t_0$ with $\varphi(t_1) > \frac{1}{2}(\varphi(t_0)+\varphi^{**}(t_0))$. Let $k := \frac{\varphi(t_0)-\varphi(t_1)}{t_0-t_1}$. Since φ is increasing, it follows by convexity that

$$\varphi(t) \geqslant k(t - t_0) + \tfrac{1}{2}(\varphi(t_0) + \varphi^{**}(t_0)).$$

Therefore, by Young's inequality for φ^*,

$$\varphi^*(k) = \sup_{t \geqslant 0} \left(kt - \varphi(t)\right) \leqslant kt_0 - \tfrac{1}{2}(\varphi(t_0) + \varphi^{**}(t_0))$$

$$\leqslant \varphi^*(k) + \varphi^{**}(t_0) - \tfrac{1}{2}(\varphi(t_0) + \varphi^{**}(t_0)) < \varphi^*(k),$$

a contradiction. The case $\varphi(t_0) = \infty$ is handled similarly, with the estimate $\varphi(t) \geqslant k(t - t_\infty) + 2\varphi^{**}(t_0)$ for suitably big k.

We next consider the general case $\varphi \in \Phi_{\mathrm{w}}$. By Theorem 2.2.3, there exists $\psi \in \Phi_{\mathrm{s}}$ with $\varphi \simeq \psi$. Using Lemma 2.4.3(c) twice we obtain $\varphi^{**} \simeq \psi^{**}$. By the first part of the proof, $\psi = \psi^{**}$ and thus $\varphi^{**} \simeq \psi \simeq \varphi$.

We already know by (2.4.2) that φ^{**} is a convex minorant of φ. Suppose that ψ is also a convex minorant of φ. By taking $\max\{\psi, 0\}$ we may assume that ψ is non-negative. By Lemma 2.4.3(a), used twice, we have $\psi^{**} \leqslant \varphi^{**}$. But since ψ is convex, the first part of the proof implies that $\psi^{**} = \psi$ so that $\psi \leqslant \varphi^{**}$. Hence φ^{**} is the greatest convex minorant. □

Corollary 2.4.6 *Let* $\varphi, \psi \in \Phi_{\mathrm{c}}$. *Then* $\varphi \leqslant \psi$ *if and only if* $\psi^* \leqslant \varphi^*$.

Proof Lemma 2.4.3(a) yields the implication from Φ-functions to conjugate functions. The reverse implication follows using Lemma 2.4.3(a) for φ^* and ψ^* and Proposition 2.4.5 that gives $\varphi^{**} = \varphi$ and $\psi^{**} = \psi$. □

Lemma 2.4.7 *Let* $\varphi \in \Phi_{\mathrm{c}}$ *and* $b := \lim\limits_{t \to 0^+} \frac{\varphi(t)}{t} = \varphi'(0)$. *Then* $\varphi^*(s) = 0$ *if and only if* $s \leqslant b$.

Here $\varphi'(0)$ is the right derivative of a convex function at the origin.

Proof Since φ is convex, it satisfies (Inc)$_1$. In particular, $\frac{\varphi(t)}{t} \geqslant \varphi'(0) = b$. We observe that

$$\varphi^*(s) = \sup_{t \geqslant 0} t\left(s - \frac{\varphi(t)}{t}\right).$$

If $s \leqslant b$, then the parenthesis is non-positive, so $\varphi^*(s) = 0$. If $s > b$, then the parenthesis is positive for some sufficiently small $t > 0$, and so $\varphi^*(s) > 0$. □

The following theorem gives a simple formula for approximating the inverse of φ^*. This results was previously shown for N-functions in [34, Lemma 2.6.11] and for generalized Φ-functions in [29, Lemma 2.3] but the later proof includes a mistake.

Theorem 2.4.8 *If* $\varphi \in \Phi_{\mathrm{w}}$, *then* $\varphi^{-1}(t)(\varphi^*)^{-1}(t) \approx t$.

Proof By Theorem 2.3.6 and Lemma 2.4.3(c), the claim is invariant under equivalence of Φ-functions, so by Theorem 2.2.3 we may assume that $\varphi \in \Phi_{\mathrm{s}}$.

The convexity of φ implies (Inc)$_1$. If we combine this with the definition of the conjugate function, we get, for $s > 0$, that

$$\varphi^*\Big(\tfrac{\varphi(s)}{s}\Big) = \sup_{t \geqslant 0} t\Big(\tfrac{\varphi(s)}{s} - \tfrac{\varphi(t)}{t}\Big) = \sup_{t \in [0,s]} t\Big(\tfrac{\varphi(s)}{s} - \tfrac{\varphi(t)}{t}\Big) \leqslant \sup_{t \in [0,s]} t\tfrac{\varphi(s)}{s} = \varphi(s).$$

On the other hand, choosing $t = s$ in the supremum, we find that

$$\varphi^*\Big(2\tfrac{\varphi(s)}{s}\Big) = \sup_{t \geqslant 0} \Big(2\tfrac{\varphi(s)}{s}t - \varphi(t)\Big) \geqslant 2\varphi(s) - \varphi(s) = \varphi(s).$$

Thus we have shown that

$$\varphi^*\Big(\tfrac{\varphi(s)}{s}\Big) \leqslant \varphi(s) \leqslant \varphi^*\Big(2\tfrac{\varphi(s)}{s}\Big).$$

The claim of the lemma is immediate for $t = 0$, so we may assume that $0 < t < \infty$. In this case we can find $s > 0$ such that $t = \varphi(s)$ and $\varphi^{-1}(t) = s$ since $\varphi \in \Phi_{\mathrm{s}}$; then $\tfrac{\varphi(s)}{s} = \tfrac{t}{\varphi^{-1}(t)} =: u$ and the previous inequality can be written as

$$\varphi^*(u) \leqslant t \leqslant \varphi^*(2u). \tag{2.4.3}$$

By Lemma 2.3.9, $(\varphi^*)^{-1}$ is increasing and $(\varphi^*)^{-1}(\varphi^*(2u)) \leqslant 2u$. We obtain

$$(\varphi^*)^{-1}(t) \leqslant (\varphi^*)^{-1}\big(\varphi^*(2u)\big) \leqslant 2u = \frac{2t}{\varphi^{-1}(t)},$$

so that $\varphi^{-1}(t)(\varphi^*)^{-1}(t) \leqslant 2t$.

It remains to prove the lower bound. We assume first that $\varphi'(0) = 0$, where $\varphi'(0)$ is the right-derivative of a convex function at the origin. Lemma 2.4.7 implies that $\varphi^*(v) > 0$ for all $v > 0$. Then Lemma 2.3.3 gives that

$$(\varphi^*)^{-1}\big(\varphi^*(u)\big) = \min\big\{u, \varphi^*(u_\infty^*)\big\},$$

where $u_\infty^* := \inf\{v : \varphi^*(v) = \infty\}$. If $u \leqslant \varphi^*(u_\infty^*)$, then this and (2.4.3) imply the desired lower bound: $\tfrac{t}{\varphi^{-1}(t)} = u \leqslant (\varphi^*)^{-1}(t)$. If $u > \varphi^*(u_\infty^*)$, then $\infty = \varphi^*(u) \leqslant t < \infty$ by (2.4.3) and by the choice of t this case cannot occur.

We have proved the claim in the case $\varphi'(0) = 0$. Suppose then that $\varphi'(0) =: b > 0$. Let us define $\psi(t) := \varphi(t) - bt$. Straight from the definition we find that ψ is convex, and so $\tfrac{\psi(t)}{t}$ is increasing. Thus $\psi(t) = t\big(\tfrac{\varphi(t)}{t} - b\big)$ belongs to Φ_{c} and moreover $\psi'(0) = 0$. Since $\varphi(t) \leqslant 2\max\{bt, \psi(t)\}$, we obtain $\varphi^{-1}(s) \geqslant \tfrac{1}{2}\min\{\tfrac{s}{b}, \psi^{-1}(s)\}$ by Lemma 2.3.8. On the other hand,

$$\varphi^*(s) = \sup_{t \geqslant 0}((s-b)t - \psi(t)) = \psi^*\big(\max\{s-b, 0\}\big).$$

Hence $(\varphi^*)^{-1}(t) = b + (\psi^*)^{-1}(t)$ for $t > 0$. With these expressions we calculate that

$$\varphi^{-1}(t)(\varphi^*)^{-1}(t) \geqslant \tfrac{1}{2} \min\{\tfrac{t}{b}, \psi^{-1}(t)\}\big(b + (\psi^*)^{-1}(t)\big)$$

$$\geqslant \tfrac{1}{2} \min\{t, \psi^{-1}(t)(\psi^*)^{-1}(t)\}.$$

By the first part of the proof, $\psi^{-1}(t)(\psi^*)^{-1}(t) \gtrsim t$ since $\psi'(0) = 0$. Thus we have proved the lower bound also in the case $\varphi'(0) > 0$. $\qquad\square$

As part of the previous proof, we found that

$$\varphi^*\left(\tfrac{\varphi(t)}{t}\right) \leqslant \varphi(t)$$

for convex Φ-functions. If $\varphi \in C^1$, then $(\mathrm{Inc})_1$ is equivalent to the inequality $\frac{\varphi(t)}{t} \leqslant \varphi'(t)$, since the derivative of $\frac{\varphi(t)}{t}$ is non-negative. In this case we can derive the following variant of the previous inequality (from [61, Proposition 3.6(4)]):

$$\varphi^*(x, \varphi'(x, t)) \leqslant t\varphi'(x, t). \tag{2.4.4}$$

The proof is as follows. Since φ is convex, $\varphi(x, s) \geqslant \varphi(x, t) + k(s - t)$, where $k := \varphi'(x, t)$ is the slope. Then from the definition of the conjugate function we have

$$\varphi^*(x, \varphi'(x, t)) = \sup_{s \geqslant 0}(sk - \varphi(x, s))$$

$$\leqslant \sup_{s \geqslant 0}(sk - \varphi(x, t) - k(s - t)) = tk - \varphi(x, t) \leqslant t\varphi'(x, t).$$

The next results was previously shown for N-functions in [52, Lemma 5.2]. Note that the next result could also be derived as consequence of Theorem 2.4.8 and Proposition 2.3.7, but here we provide a direct proof which avoids the inverse function.

Proposition 2.4.9 *Let $\varphi \in \Phi_w$. Then φ satisfies $(\mathrm{aInc})_p$ or $(\mathrm{aDec})_q$ if and only if φ^* satisfies $(\mathrm{aDec})_{p'}$ or $(\mathrm{aInc})_{q'}$, respectively.*

Proof We start with the special cases $(\mathrm{Inc})_p$ and $(\mathrm{Dec})_q$. We have that φ satisfies $(\mathrm{Inc})_p$ if and only if $\frac{\varphi(t^{1/p})}{t}$ is increasing, similarly for φ^* and $(\mathrm{Dec})_q$. From the definition of the conjugate function,

$$\frac{\varphi^*(s^{\frac{1}{p}})}{s} = \frac{1}{s} \sup_{t \geqslant 0}\left(ts^{\frac{1}{p}} - \varphi(t)\right) = \sup_{v \geqslant 0} v\left(v^{-\frac{1}{p}} - \frac{\varphi\big((sv)^{\frac{1}{p'}}\big)}{sv}\right),$$

where we used the change of variables $t =: (sv)^{\frac{1}{p'}}$. From this expression, we see that φ^* satisfies $(\text{Dec})_q$ and $(\text{Inc})_p$ if φ satisfies $(\text{Inc})_{q'}$ and $(\text{Dec})_{p'}$, respectively. For the opposite implication, we use $(\varphi^*)^* \simeq \varphi$ from Proposition 2.4.5.

Suppose now that φ satisfies $(\text{aInc})_p$. Then $\psi(s) := s^p \inf_{t \geqslant s} t^{-p} \varphi(t)$ satisfies $(\text{Inc})_p$. A short calculation shows that $\varphi \approx \psi$: by (aInc) we obtain

$$\tfrac{1}{a} \varphi(s) = \tfrac{1}{a} s^p \frac{\varphi(s)}{s^p} \leqslant \psi(s) \leqslant s^p \frac{\varphi(s)}{s^p} = \varphi(s).$$

By the above argument, ψ^* satisfies $(\text{Dec})_q$ and by Lemma 2.4.3(c), $\varphi^* \simeq \psi^*$. For $(\text{aDec})_q$, we can argue in the same way with the auxiliary function $\psi(s) := s^q \sup_{t \leqslant s} t^{-q} \varphi(t)$. □

Definition 2.4.10 We say that $\varphi \in \Phi_w$ satisfies ∇_2, if φ^* satisfies Δ_2.

We can now connect this concept from the theory of Orlicz spaces to our assumptions as Proposition 2.4.9 and Lemma 2.2.6 yield the following result.

Corollary 2.4.11 *A function $\varphi \in \Phi_w$ satisfies ∇_2 if and only if it satisfies* (aInc).

2.5　Generalized Φ-Functions

Next we generalize Φ-functions in such a way that they may depend on the space variable. Let (A, Σ, μ) be a σ-finite, complete measure space. In what follows we always make the natural assumption that our measure μ is not identically zero.

Definition 2.5.1 Let $f : A \times [0, \infty) \to \mathbb{R}$ and $p, q > 0$. We say that f satisfies $(\text{aInc})_p$ or $(\text{aDec})_q$, if there exists $a \geqslant 1$ such that the function $t \mapsto f(x, t)$ satisfies $(\text{aInc})_p$ or $(\text{aDec})_q$ with a constant a, respectively, for μ-almost every $x \in A$. When $a = 1$, we use the notation $(\text{Inc})_p$ and $(\text{Dec})_q$.

Note that in the almost increasing and decreasing conditions we require that the same constant applies to almost every point. Furthermore, if we define $f(x, t) = g(t)$ for every x, then f satisfies (aInc) in the sense of the previous definition if and only if g satisfies (aInc) in the sense of Definition 2.1.2. The same applies to the other terms. Therefore, there is no need to distinguish between the conditions based on whether there is an x-dependence of the function or not.

Definition 2.5.2 Let (A, Σ, μ) be a σ-finite, complete measure space. A function $\varphi : A \times [0, \infty) \to [0, \infty]$ is said to be a (*generalized*) Φ-*prefunction* on (A, Σ, μ) if $x \mapsto \varphi(x, |f(x)|)$ is measurable for every $f \in L^0(A, \mu)$ and $\varphi(x, \cdot)$ is a Φ-prefunction for μ-almost every $x \in A$. We say that the Φ-prefunction φ is

- a (*generalized weak*) Φ-*function* if φ satisfies $(\text{aInc})_1$;
- a (*generalized*) convex Φ-*function* if $\varphi(x, \cdot) \in \Phi_c$ for μ-almost all $x \in A$;
- a (*generalized*) strong Φ-*function* if $\varphi(x, \cdot) \in \Phi_s$ for μ-almost all $x \in A$.

If φ is a generalized weak Φ-function on (A, Σ, μ), we write $\varphi \in \Phi_w(A, \mu)$ and similarly we define $\varphi \in \Phi_c(A, \mu)$ and $\varphi \in \Phi_s(A, \mu)$. If Ω is an open subset of $\mathbb{R}^n$ and μ is the n-dimensional Lebesgue measure we omit μ and abbreviate $\Phi_w(\Omega)$, $\Phi_c(\Omega)$ or $\Phi_s(\Omega)$. Or we say that φ is a generalized (weak/convex/strong) Φ-function on Ω. Unless there is danger of confusion, we will omit the word "generalized".

Clearly $\Phi_s(A, \mu) \subset \Phi_c(A, \mu) \subset \Phi_w(A, \mu)$. Every Φ-function is a generalized Φ-function if we set $\varphi(x, t) := \varphi(t)$ for $x \in A$ and $t \in [0, \infty)$. Next we give some examples of non-trivial generalized Φ-functions.

Example 2.5.3 Let $p : A \to [1, \infty]$ be a measurable function and define $p_\infty := \limsup_{|x| \to \infty} p(x)$. Let us interpret $t^\infty := \infty \chi_{(1,\infty]}(t)$. Let $a : A \to (0, \infty)$ be a measurable function and $1 \leqslant r < q < \infty$. Let us define, for $t \geqslant 0$,

$$\varphi_1(x, t) := t^{p(x)} a(x),$$

$$\varphi_2(x, t) := t^{p(x)} \log(e + t),$$

$$\varphi_3(x, t) := \min\{t^{p(x)}, t^{p_\infty}\},$$

$$\varphi_4(x, t) := t^{p(x)} + \sin(t),$$

$$\psi_1(x, t) := t^p + a(x)t^q,$$

$$\psi_2(x, t) := (t - 1)^r_+ + a(x)(t - 1)^q_+.$$

We observe that

- $\varphi_3 \in \Phi_w(A, \mu) \setminus \Phi_c(A, \mu)$ when p is non-constant,
- $\varphi_4 \in \Phi_w(A, \mu) \setminus \Phi_c(A, \mu)$ when $\inf_{x \in A} p(x) \leqslant \frac{3}{2}$,
- $\varphi_1, \varphi_2 \in \Phi_c(A, \mu) \setminus \Phi_s(A, \mu)$ when $p = \infty$ in a set of positive measure, and
- $\psi_1, \psi_2 \in \Phi_s(A, \mu)$ when $p, q \in [1, \infty)$.

Moreover, if $p(x) < \infty$ for μ-almost every x, then $\varphi_1, \varphi_2 \in \Phi_s(A, \mu)$.

Measurability

Note that in the definition of generalized Φ-functions we have directly assumed that $x \mapsto \varphi(x, |f(x)|)$ is measurable. If φ is left-continuous, then this assumption can be replaced with the conditions from the next theorem.

Theorem 2.5.4 *Let* $\varphi \colon A \times [0, \infty) \to [0, \infty]$, $x \mapsto \varphi(x, t)$ *be measurable for every* $t \geqslant 0$ *and* $t \mapsto \varphi(x, t)$ *be increasing and left-continuous for* μ-*almost every* x. *If* $f \in L^0(A, \mu)$ *is measurable, then* $x \mapsto \varphi(x, |f(x)|)$ *is measurable.*

Proof We have to show that $E_a := \{x \in A : \varphi(x, |f(x)|) > a\}$ is measurable for every $a \in \mathbb{R}$. Let us write $F_a(t) := \{x \in A : \varphi(x, t) > a\} \cap \{x \in A : |f(x)| \geqslant t\}$, for $t \geqslant 0$. Then for each t we have $F_a(t) \subset E_a$ since φ is increasing. Assume then that $x \in E_a$. Let (t_j) be a sequence of non-negative rational numbers converging to $|f(x)|$ from below. By the left-continuity of φ, we have $\lim_{j \to \infty} \varphi(x, t_j) = \varphi(x, |f(x)|)$. Thus there exists j_0 such that $\varphi(x, t_{j_0}) > a$ and $0 \leqslant t_{j_0} \leqslant |f(x)|$. This yields that $x \in F_a(t_{j_0})$. We have shown $E_a = \bigcup_{t \in \mathbb{Q} \cap [0, \infty)} F_a(t)$. Since each $F_a(t)$ is measurable by assumption and the union is countable, the set E_a is measurable.   □

The next example shows that $x \mapsto \varphi(x, |f(x)|)$ need not to be measurable if we omit left-continuity of φ.

Example 2.5.5 Consider the Lebesgue measure on $[1, 2]$ and let $F \subset [1, 2]$ be a non-measurable set. We define $\varphi : [1, 2] \times [0, \infty) \to [0, \infty]$ by

$$\varphi(x, t) := \chi_F(t)\chi_{\{x\}}(t) + \infty \chi_{(x, \infty)}(t).$$

For constant $t \geqslant 0$, $x \mapsto \varphi(x, t)$ is decreasing and hence measurable. For each $x \in [1, 2]$, $t \mapsto \varphi(x, t)$ belongs to Φ_w, but it is left-continuous only when $\chi_F(x) = 0$ i.e. when $x \notin F$. Let $f : [1, 2] \to \mathbb{R}$, $f(x) := x$. Then f is continuous, and hence measurable. But $\varphi(x, |f(x)|) = \varphi(x, x) = \chi_F(x)$ is not a measurable function.

Properties of Φ-functions are generalized point-wise uniformly to the generalized Φ-function case. For instance we define equivalence as follows.

Definition 2.5.6 We say that $\varphi, \psi : A \times [0, \infty) \to [0, \infty]$ are *equivalent*, $\varphi \simeq \psi$, if there exist $L > 1$ such that for all $t \geqslant 0$ and μ-almost all $x \in A$ we have

$$\psi(x, \tfrac{t}{L}) \leqslant \varphi(x, t) \leqslant \psi(x, Lt).$$

The above definitions have been chosen so all the results from Sects. 2.1–2.4 can be adapted to generalized (weak/convex/strong) Φ-functions. Most of the results are point-wise and thus can be used in the generalized case without problems. The issue is that equivalence does not ensure measurability with respect to the first variable: consider for example φ from Example 2.5.5 and the equivalent, left-continuous Φ-function φ_∞. Results which assert the existence of a new Φ-function are Lemma 2.1.8(a), Lemma 2.2.1, Theorem 2.2.3, Proposition 2.2.7, Proposition 2.3.13 and Lemma 2.4.2. Next we show that the proofs have been chosen so that they preserve measurability. We have collected the correspondence between results in Table 2.1.

Let us start from Lemma 2.1.8, where the crucial point is that measurability is placed as an *a priori* assumption for both φ and ψ, rather than being a consequence in (a). With this modification, the proof remains the same.

Lemma 2.5.7 *Let $\varphi, \psi : A \times [0, \infty) \to [0, \infty]$, $\varphi \simeq \psi$, be increasing with respect to the second variable, and $x \mapsto \varphi(x, |f(x)|)$ and $x \mapsto \psi(x, |f(x)|)$ be measurable for every measurable f.*

(a) *If φ is a generalized Φ-prefunction, then ψ is a generalized Φ-prefunction.*
(b) *If φ satisfies $(aInc)_p$, then ψ satisfies $(aInc)_p$.*
(c) *If φ satisfies $(aDec)_q$, then ψ satisfies $(aDec)_q$.*

Table 2.1 Correspondence bewteen results for Φ-functions (column Φ_w) and generalized Φ-functions (column $\Phi_w(A, \mu)$)

	Φ_w	$\Phi_w(A, \mu)$
Equivalence of Φ-prefunctions	Lemma 2.1.8	Lemma 2.5.7
Equivalent convex function	Lemma 2.2.1	Lemma 2.5.9
Equivalent strong Φ-function	Theorem 2.2.3	Theorem 2.5.10
Equivalent strong bijection	Proposition 2.2.7	Proposition 2.5.11
Inverse is Φ-function	Proposition 2.3.13	Proposition 2.5.14
Conjugate is Φ-function	Lemma 2.4.2	Lemma 2.5.8

Let us here show how the upgrading results can be conveniently obtained by means of the conjugate function. We first show that φ^* is measurable, i.e. we generalize Lemma 2.4.2.

Lemma 2.5.8 *If $\varphi \in \Phi_w(A, \mu)$, then $\varphi^* \in \Phi_c(A, \mu)$.*

Proof By Lemma 2.4.2 and Theorem 2.5.4, it is enough to show that $x \mapsto \varphi^*(x, t)$ is measurable for every $t \geqslant 0$. We first show that

$$\sup_{u \geqslant 0}(ut - \varphi(x, u)) = \sup_{u \in \mathbb{Q} \cap [0, \infty)} (ut - \varphi(x, u)).$$

The inequality "$\geqslant$" is obvious. Suppose that $u \in (0, \infty) \setminus \mathbb{Q}$ and let $u_i \in (u - \frac{1}{i}, u) \cap \mathbb{Q}$. Since φ is increasing, we obtain $ut - \varphi(x, u) \leqslant ut - \varphi(x, u_i) \leqslant u_i t - \varphi(x, u_i) + \frac{t}{i}$. When $i \to \infty$, we obtain the inequality "$\leqslant$".

Let $a \geqslant 0$. Then $\varphi^*(x, t) \leqslant a$ if and only if $ut - \varphi(x, u) \leqslant a$ for all $u \in \mathbb{Q} \cap [0, \infty)$. Thus

$$\{x : \varphi^*(x, t) \leqslant a\} = \bigcap_{u \in \mathbb{Q} \cap [0, \infty)} \{x : ut - \varphi(x, u) \leqslant a\}$$

is measurable as a countable intersection of measurable sets, and hence $x \mapsto \varphi^*(x, t)$ is measurable. $\qquad\square$

Let us then consider Lemma 2.2.1 and Theorem 2.2.3 which show that every weak Φ-function is equivalent to a strong Φ-function.

Lemma 2.5.9 *If $\varphi \in \Phi_w(A, \mu)$ satisfies (aInc)$_p$ with $p \geqslant 1$, then there exists $\psi \in \Phi_c(A, \mu)$ equivalent to φ such that $\psi^{1/p}$ is convex. In particular, ψ satisfies (Inc)$_p$.*

Proof We first observe that $\xi := \varphi^{\frac{1}{p}} \in \Phi_w(A, \mu)$. It follows by Lemma 2.5.8 that $\xi^{**} \in \Phi_c(A, \mu)$ and by Proposition 2.4.5 that $\xi \simeq \xi^{**}$. Then $\psi := (\xi^{**})^p$ is the required convex Φ-function. $\qquad\square$

Theorem 2.5.10 *Every weak Φ-function is equivalent to a strong Φ-function.*

Proof Let φ_c be from Lemma 2.5.9. By the proof of Theorem 2.2.3 we need only to show that the functions in the proof satisfy the measurability property.

There we defined $t_\infty(x) := \inf\{t \; : \; \varphi_c(x,t) = \infty\} \in (0,\infty)$ and

$$\varphi_s(x,t) := \varphi_c(x,t) + \frac{2t - t_\infty(x)}{t_\infty(x) - t} \chi_{(\frac{1}{2}t_\infty(x),t_\infty(x))}(t) + \infty \chi_{[t_\infty(x),\infty)}(t).$$

Then φ_s is left-continuous and hence by Theorem 2.5.4 we need to show that $x \mapsto \varphi_s(x,t)$ is measurable for every $t \geqslant 0$.

This is clear if $x \mapsto t_\infty(x)$ is measurable. Let $a \geqslant 0$. Then $\inf\{t \; : \; \varphi_c(x,t) = \infty\} \leqslant a$ if and only if $\varphi_c(x, a + \varepsilon) = \infty$ for all $\varepsilon > 0$. The later implies that $\varphi_c(x, a + r) = \infty$ for all $r \in \mathbb{Q} \cap (0, \infty)$. We obtain that

$$\{x \; : \; t_\infty(x) \leqslant a\} = \bigcap_{r \in \mathbb{Q} \cap (0,\infty)} \{x \; : \; \varphi(x, a + r) = \infty\}$$

is measurable as a countable intersection of measurable sets, and hence t_∞ is measurable.   □

The proof of following proposition is the same as the proof of Proposition 2.2.7 except that it is based on Lemma 2.5.9, not its preliminary version Lemma 2.2.1.

Proposition 2.5.11 *If* $\varphi \in \Phi_w(A, \mu)$ *satisfies* (aDec), *then there exists* $\psi \in \Phi_s(A, \mu)$ *with* $\psi \approx \varphi$ *such that* $t \mapsto \psi(x,t)$ *is a strictly increasing bijection for μ-almost every* $x \in A$.

Then we consider the inverse of a generalized Φ-function. For that we prove an extra lemma.

Lemma 2.5.12 *Let* $\varphi \colon A \times [0, \infty) \to [0, \infty]$. *If* $t \mapsto \varphi(x,t)$ *is increasing for μ-almost every* x *and if* $x \mapsto \varphi(x,t)$ *is measurable for every* $t \geqslant 0$, *then* $x \mapsto \varphi^{-1}(x, |f(x)|)$ *is measurable for every measurable* f.

Proof By Lemma 2.3.9(c), φ^{-1} is left-continuous and hence by Theorem 2.5.4 we need to show that $x \mapsto \varphi^{-1}(x, \tau)$ is measurable for every $\tau \geqslant 0$.

Let $a, \tau \geqslant 0$. Then $\varphi^{-1}(x, \tau) = \inf\{t \; : \; \varphi(x,t) \geqslant \tau\} > a$ if and only if there exists $\varepsilon > 0$ such that $\varphi(x, a + \varepsilon) < \tau$. Since φ is increasing with respect to the second variable, the later implies that $\varphi(x, a + r) < \tau$ for all $r \in \mathbb{Q} \cap (0, \varepsilon]$. Thus

$$\{x \; : \; \varphi^{-1}(x, \tau) > a\} = \bigcup_{r \in \mathbb{Q} \cap (0,\infty)} \{x \; : \; \varphi(x, a + r) < \tau\}$$

is measurable as a countable union of measurable sets, and hence $x \mapsto \varphi^{-1}(x, \tau)$ is measurable.   □

Next we generalize Definition 2.3.12 to $\Phi_w(A, \mu)$-functions.

Definition 2.5.13 We say that $\xi : A \times [0, \infty] \to [0, \infty]$ belongs to $\Phi_{\mathrm{w}}^{-1}(A, \mu)$ if it satisfies (aDec)$_1$, $x \mapsto \xi(x, t)$ is measurable for all t and if for μ-almost every $x \in A$ the function $t \mapsto \xi(x, t)$ is increasing, left-continuous, and $\xi(x, t) = 0$ if and only if $t = 0$ and $\xi(x, t) = \infty$ if and only if $t = \infty$.

Lemma 2.5.12 and Proposition 2.3.13 show that this class characterizes inverses of left-continuous generalized weak Φ-functions and that $^{-1}$ is an involution on the set of left-continuous generalized weak Φ-functions. We denote be $\Phi_{\mathrm{w}+}(A, \mu)$ the set of left-continuous generalized weak Φ-functions.

Proposition 2.5.14 *The transformation $\varphi \mapsto \varphi^{-1}$ is a bijection from $\Phi_{\mathrm{w}+}$ to Φ_{w}^{-1}:*

(a) *If $\varphi \in \Phi_{\mathrm{w}+}(A, \mu)$, then $\varphi^{-1} \in \Phi_{\mathrm{w}}^{-1}(A, \mu)$ and $(\varphi^{-1})^{-1} = \varphi$.*
(b) *If $\xi \in \Phi_{\mathrm{w}}^{-1}(A, \mu)$, then $\xi^{-1} \in \Phi_{\mathrm{w}+}(A, \mu))$ and $(\xi^{-1})^{-1} = \xi$.*

Limits of Weak Φ-Functions

In this subsection, we consider two types of limits of functions in $\Phi_{\mathrm{w}}(\Omega)$, $\Omega \subset \mathbb{R}^n$, with respect to x which are used in Chap. 4 in the conditions (A1) and (A2). In each case we end up with a function $\tilde{\varphi}$ in Φ_{w} independent of x. Note that there is never any problem with measurability in this case, since $\tilde{\varphi}$ is increasing and hence $\{x : \tilde{\varphi}(|g(x)|) \geqslant a\}$ is either $\{x : |g(x)| > \tilde{\varphi}^{-1}(a)\}$ or $\{x : |g(x)| \geqslant \tilde{\varphi}^{-1}(a)\}$.

For $\varphi \in \Phi_{\mathrm{w}}(\Omega)$ and $B \subset \mathbb{R}^n$, we define

$$\varphi_B^+(t) := \operatorname*{ess\,sup}_{x \in B \cap \Omega} \varphi(x, t) \quad \text{and} \quad \varphi_B^-(t) := \operatorname*{ess\,inf}_{x \in B \cap \Omega} \varphi(x, t).$$

Example 2.5.15 Let $\varphi_1(x, t) := t^2 |x|$ and $\varphi_2(x, t) := t^2 \frac{1}{|x|}$. Then $\varphi_1, \varphi_2 \in \Phi_{\mathrm{s}}(\mathbb{R}^n)$ but $(\varphi_1)_B^- \equiv 0$ and $(\varphi_2)_B^+ \equiv \infty$ for every ball B containing the origin.

We say that $\varphi \in \Phi_{\mathrm{w}}$ is *degenerate* if $\varphi|_{(0,\infty)} \equiv 0$ or $\varphi|_{(0,\infty)} \equiv \infty$.

Lemma 2.5.16 *Let $\varphi \in \Phi_{\mathrm{w}}(B)$. If φ_B^+ or φ_B^- is non-degenerate, then it is a weak Φ-function.*

Proof We only consider φ_B^-, as the proof for φ_B^+ is analogous. Clearly $\varphi_B^-(0) = 0$ and $\lim_{t \to 0^+} \varphi_B^-(t) \leqslant \lim_{t \to 0^+} \varphi(x, t) = 0$. Let $0 < s < t$ and $\varepsilon > 0$. Choose $x \in B$ such that $\varphi(x, t) \leqslant \varphi_B^-(t) + \varepsilon$. Then $\varphi_B^-(s) \leqslant \varphi(x, s) \leqslant \varphi(x, t) \leqslant \varphi_B^-(t) + \varepsilon$ and by letting $\varepsilon \to 0^+$ we see that φ_B^- is increasing. Similarly

$$\frac{\varphi_B^-(s)}{s} \leqslant \frac{\varphi(x, s)}{s} \leqslant a \frac{\varphi(x, t)}{t} \leqslant a \frac{\varphi_B^-(t) + \varepsilon}{t}$$

and (aInc)$_1$ follows by letting $\varepsilon \to 0^+$. If φ_B^- is not degenerate, then (aInc)$_1$ implies that $\lim_{t\to\infty} \varphi_B^-(t) = \infty$.   □

In fact, if φ in the previous lemma is left-continuous, then the same is true for φ_B^+, as well: Let $0 < s < t$ and $\varepsilon > 0$. We choose $x \in B$ such that $\varphi(x,t) \geqslant \varphi_B^+(t) - \varepsilon$. Then

$$\liminf_{s\to t^-} \varphi_B^+(s) \geqslant \liminf_{s\to t^-} \varphi(x,s) = \varphi(x,t) \geqslant \varphi_B^+(t) - \varepsilon$$

and letting $\varepsilon \to 0^+$ we obtain $\liminf_{s\to t^-} \varphi_B^+(s) \geqslant \varphi_B^+(t)$. Since φ_B^+ is increasing we have $\limsup_{s\to t^-} \varphi_B^+(s) \leqslant \varphi_B^+(t)$.

Even if φ is left-continuous and non-degenerate, φ_B^- need not to be left-continuous. This is one of the main reasons why we did not include left-continuity as a requirement in the definition of Φ_w.

Example 2.5.17 Let $\varphi(x,t) := t^{1+1/|x|}$. Then $\varphi(x,1) = 1$ and hence φ is non-degenerate. Let $B := B(0,r)$, $r > 0$. A short calculation gives

$$\varphi_B^-(t) = \begin{cases} 0, & \text{if } t \in [0,1) \\ t^{1+1/r}, & \text{if } t \geqslant 1 \end{cases}$$

and

$$\varphi_B^+(t) = \begin{cases} t^{1+1/r}, & \text{if } t \in [0,1] \\ \infty, & \text{if } t > 1. \end{cases}$$

Thus φ_B^- is not left-continuous at 1.

The second limit that we consider is the behavior "at infinity". For this we define some asymptotical Φ-functions:

$$\varphi_\infty^+(t) := \limsup_{|x|\to\infty} \varphi(x,t) \quad \text{and} \quad \varphi_\infty^-(t) := \liminf_{|x|\to\infty} \varphi(x,t).$$

If $\varphi \simeq \psi$, then a short calculation gives $\varphi_\infty^\pm \simeq \psi_\infty^\pm$.

Note that $\varphi_\infty^\pm$ can be degenerate. For instance, if $\varphi(x,t) := t^2|x|^{-1}$, then $\varphi_\infty^\pm(t) = 0$ for every $t \geqslant 0$, and if $\varphi(x,t) := t(1+|x|)$, then $\varphi_\infty^\pm(t) = \infty$ for every $t > 0$. However, if this is not the case, then it is a weak Φ-function:

Lemma 2.5.18 *If $\varphi \in \Phi_w(\Omega)$ and φ_∞^+ or φ_∞^- is non-degenerate, then it is a weak Φ-function.*

Proof Note first that $\varphi_\infty^+(0) = \limsup_{|x|\to\infty} \varphi(x,0) = 0$. Let $0 < s < t$. Since φ is increasing we have $\varphi(x,s) \leqslant \varphi(x,t)$ and hence $\limsup_{|x|\to\infty} \varphi(x,s) \leqslant$

$\limsup_{|x|\to\infty} \varphi(x,t)$ so φ_∞ is increasing. Since φ satisfies $(\text{aInc})_1$ we obtain

$$\limsup_{|x|\to\infty} \frac{\varphi(x,s)}{s} \lesssim \limsup_{|x|\to\infty} \frac{\varphi(x,t)}{t}$$

and hence φ_∞^+ satisfies $(\text{aInc})_1$. If we choose $t > 0$ with $\varphi_\infty^+(t) < \infty$, then this yields $\lim_{s\to 0} \varphi_\infty^+(s) = 0$. Conversely, $s > 0$ with $\varphi_\infty^+(s) > 0$ yields that $\lim_{t\to\infty} \varphi_\infty^+(t) = \infty$. Such s and t exist since φ_∞^+ is non-degenerate. Thus $\varphi_\infty^+ \in \Phi_{\text{w}}$. The proof for φ_∞^- is analogous. $\qquad\qquad\square$

Moreover, if φ in the previous lemma is convex, then so is φ_∞^+, as we now show. Let $r > 0$, $t, s \geqslant 0$ and $\theta \in [0, 1]$. We obtain

$$\sup_{|x|>r} \left\{\varphi(x, \theta t + (1-\theta)s)\right\} \leqslant \sup_{|x|>r} \left\{\theta\varphi(x,t) + (1-\theta)\varphi(x,s)\right\}$$

$$\leqslant \sup_{|x|>r}\left\{\theta\varphi(x,t)\right\} + \sup_{|x|>r}\left\{(1-\theta)\varphi(x,s)\right\}$$

$$= \theta \sup_{|x|>r}\left\{\varphi(x,t)\right\} + (1-\theta)\sup_{|x|>r}\left\{\varphi(x,s)\right\}$$

and by taking $r \to \infty$ we obtain $\varphi_\infty^+(\theta t + (1-\theta)s) \leqslant \theta\varphi_\infty^+(t) + (1-\theta)\varphi_\infty^+(s)$. Hence φ_∞^+ is convex.

Example 2.5.19 Let $\varphi(x,t) = t^{1+|x|}$. Then $\varphi \in \Phi_{\text{s}}(\mathbb{R}^n)$. A short calculation gives

$$\varphi_\infty^+(t) = \varphi_\infty^-(t) = \begin{cases} 0, & \text{if } t \in [0,1) \\ 1, & \text{if } t = 1 \\ \infty & \text{if } t > 1 \end{cases}$$

Then φ_∞^+ and φ_∞^- are not left-continuous at 1 and hence $\varphi_\infty^+, \varphi_\infty^- \in \Phi_{\text{w}} \setminus \Phi_{\text{s}}$ even though $\varphi \in \Phi_{\text{s}}(\Omega)$.

Weak Equivalence and Weak Doubling

We can also define some properties which are properly generalized in the sense that they have no analogue in the case that φ does not depend on the space variable.

Definition 2.5.20 We say that $\varphi, \psi : A \times [0, \infty) \to [0, \infty]$ are *weakly equivalent*, $\varphi \sim \psi$, if there exist $L > 1$ and $h \in L^1(A, \mu)$ such that

$$\varphi(x,t) \leqslant \psi(x, Lt) + h(x) \quad \text{and} \quad \psi(x,t) \leqslant \varphi(x, Lt) + h(x)$$

for all $t \geqslant 0$ and μ-almost all $x \in A$.

An easy calculation shows that $\sim$ is an equivalence relation. It clear from the definitions that $\varphi \simeq \psi$ implies $\varphi \sim \psi$ (with $h = 0$). Later in Theorem 3.2.6 we show that $\varphi \sim \psi$ if and only if $L^\varphi(A, \mu) = L^\psi(A, \mu)$. Also weak equivalence is preserved under conjugation:

Lemma 2.5.21 *Let* $\varphi, \psi : A \times [0, \infty) \to [0, \infty]$. *If* $\varphi \sim \psi$, *then* $\varphi^* \sim \psi^*$.

Proof Let $\varphi \sim \psi$. Then we obtain

$$\varphi^*(x, Lu) = \sup_{t \geqslant 0}(tLu - \varphi(x, t)) \geqslant \sup_{t \geqslant 0}(tLu - \psi(x, Lt) - h(x))$$

$$= \sup_{t \geqslant 0}(Ltu - \psi(x, Lt)) - h(x) = \psi^*(x, u) - h(x)$$

and similarly $\varphi^*(x, u) \leqslant \psi^*(x, Lu) + h(x)$.   □

Definition 2.5.22 We say that $\varphi : A \times [0, \infty) \to [0, \infty]$ satisfies the *weak doubling condition* Δ_2^{w} if there exist a constant $K \geqslant 2$ and $h \in L^1(A, \mu)$ such that

$$\varphi(x, 2t) \leqslant K\varphi(x, t) + h(x)$$

for μ-almost every $x \in A$ and all $t \geqslant 0$. We say that φ satisfies condition ∇_2^{w} if φ^* satisfies Δ_2^{w}.

If $h \equiv 0$, then we say that the (strong) Δ_2 and ∇_2 conditions hold.

Note that the Δ_2 and ∇_2-conditions for x-independent Φ-prefunctions have been defined in Definition 2.2.5. By writing $\varphi(x, t) := \varphi(t)$ we see that the definitions are equivalent. Since the constant K in Definition 2.5.22 is the same for μ-almost every $x \in A$, we see by Lemma 2.2.6 that Δ_2 is equivalent to (aDec) and by Corollary 2.4.11 that ∇_2 is equivalent to (aInc).

Lemma 2.5.23 *Let* $\varphi, \psi : A \times [0, \infty) \to [0, \infty]$ *with* $\varphi \sim \psi$.

(a) *If* φ *satisfies* Δ_2^{w}, *then* ψ *satisfies* Δ_2^{w}.
(b) *If* φ *satisfies* ∇_2^{w}, *then* ψ *satisfies* ∇_2^{w}.

Proof (a) Choose an integer $k \geqslant 1$ such that $2^{k-1} < 2L^2 \leqslant 2^k$. Then, by iterating the Δ_2^{w} assumption, we conclude that

$$\varphi(x, 2Lt) \leqslant \varphi(x, 2^k \tfrac{t}{L}) \leqslant K\varphi(x, 2^{k-1}\tfrac{t}{L}) + h(x) \leqslant \cdots \lesssim \varphi(x, \tfrac{t}{L}) + h(x).$$

Denote by h_2 the function from $\sim$. We find that

$$\psi(x, 2t) \leqslant \varphi(x, 2Lt) + h_2(x) \lesssim \varphi(x, \tfrac{t}{L}) + h(x) + h_2(x)$$

$$\leqslant \psi(x, t) + h(x) + 2h_2(x).$$

(b) Since $\varphi \sim \psi$, Lemma 2.5.21 yields $\varphi^* \sim \psi^*$. Since φ^* satisfies Δ_2^{w} so does ψ^* by (a). This means that ψ satisfies ∇_2^{w}.   □

Next we show that weak doubling can be upgraded to strong doubling via weak equivalence of Φ-functions.

Theorem 2.5.24 *If* $\varphi \in \Phi_w(A, \mu)$ *satisfies* Δ_2^w *and/or* ∇_2^w, *then there exists* $\psi \in \Phi_w(A, \mu)$ *with* $\varphi \sim \psi$ *satisfying* Δ_2 *and/or* ∇_2.

Proof By Theorem 2.5.10 and Lemmas 2.5.21 and 2.5.23, we may assume without loss of generality that $\varphi \in \Phi_s(A, \mu)$. By the assumptions,

$$\varphi(x, 2t) \leqslant K\varphi(x, t) + h(x) \quad \text{and/or} \quad \varphi^*(x, 2t) \leqslant K\varphi^*(x, t) + h(x)$$

for some $K > 2$, $h \in L^1$, $t \geqslant 0$ and μ-almost all $x \in A$. Using $\varphi = \varphi^{**}$ (Proposition 2.4.5), the definition of the conjugate Φ-function and Lemma 2.4.3(b), we rewrite the second inequality as

$$\varphi(x, 2t) = \sup_{u \geqslant 0} \left(2tu - \varphi^*(x, u)\right)$$

$$\leqslant \sup_{u \geqslant 0} \left(2tu - \tfrac{1}{K}(\varphi^*(x, 2u) - h(x))\right)$$

$$= \sup_{u \geqslant 0} \left(2tu - \tfrac{1}{K}\varphi^*(x, 2u)\right) + \tfrac{1}{K}h(x)$$

$$= \tfrac{1}{K}\varphi^{**}(x, Kt) + \tfrac{1}{K}h(x) = \tfrac{1}{K}\varphi(x, Kt) + \tfrac{1}{K}h(x).$$

Define $t_x := \varphi^{-1}(x, h(x))$ and suppose that $t > t_x$ so that $h(x) \leqslant \varphi(x, t)$ by Lemma 2.3.3. By (Inc)$_1$, we conclude that $Kh(x) \leqslant K\varphi(x, t) \leqslant \varphi(x, Kt)$. Hence in the case $t > t_x$ we have

$$\varphi(x, 2t) \leqslant (K + 1)\varphi(x, t) \quad \text{and/or} \quad \varphi(x, 2t) \leqslant \tfrac{K+1}{K^2}\varphi(x, Kt).$$

Let $q := \log_2(K + 1)$ and $p := \frac{\log(K^2/(K+1))}{\log(K/2)}$. Note that $p > 1$ since $\frac{K^2}{K+1} > \frac{K}{2}$ and $K > 2$. Divide the first inequality by $(2t)^q$ and the second one by $(2t)^p$:

$$\frac{\varphi(x, 2t)}{(2t)^q} \leqslant \frac{K+1}{2^q}\frac{\varphi(x, t)}{t^q} = \frac{\varphi(x, t)}{t^q} \quad \text{and/or}$$

$$\frac{\varphi(x, 2t)}{(2t)^p} \leqslant \frac{(K+1)K^p}{K^2 2^p}\frac{\varphi(x, Kt)}{(Kt)^p} = \frac{\varphi(x, Kt)}{(Kt)^p}.$$

Let $s > t \geqslant t_x$. Then there exists $k \in \mathbb{N}$ such that $2^k t < s \leqslant 2^{k+1}t$. Hence

$$\frac{\varphi(x, s)}{s^q} \leqslant \frac{\varphi(x, 2^{k+1}t)}{(2^k t)^q} = 2^q\frac{\varphi(x, 2^{k+1}t)}{(2^{k+1}t)^q} \leqslant 2^q\frac{\varphi(x, 2^k t)}{(2^k t)^q} \leqslant \cdots \leqslant 2^q\frac{\varphi(x, t)}{t^q},$$

so φ satisfies (aDec)$_q$ for $t \geqslant t_x$. Similarly, we find that φ satisfies (aInc)$_p$ for $t \geqslant t_x$.

Define

$$\psi(x,t) := \begin{cases} \varphi(x,t), & \text{for } t \geqslant t_x \\ c_x t^p & \text{otherwise,} \end{cases}$$

where c_x is chosen so that the ψ is continuous at t_x. Then ψ satisfies $(\text{aInc})_p$ and/or $(\text{aDec})_q$ on $[0, t_x]$ and $[t_x, \infty)$, hence on the whole real axis.

Furthermore, $\varphi(x,t) = \psi(x,t)$ when $t \geqslant t_x$, and so it follows that $|\varphi(x,t) - \psi(x,t)| \leqslant \varphi(x,t_x) \leqslant h(x)$ (Lemma 2.3.3). Since $h \in L^1$, this means that $\varphi \sim \psi$, so ψ is the required function.

Finally we show that $\psi \in \Phi_w(A, \mu)$. The function $x \mapsto c_x = \frac{\varphi(x, t_x)}{t_x^p}$ is measurable since $t_x = \varphi^{-1}(x, h(x))$ is measurable (Lemma 2.5.12), thus we obtain that $x \mapsto \psi(x, t)$ is measurable. It is clear that $t \mapsto \psi(x, t)$ is a left-continuous Φ-prefunction for μ-almost every x and hence the measurability property follows from Theorem 2.5.4. Since ψ satisfies $(\text{aInc})_1$ on $[0, t_x]$ and $[t_x, \infty)$ for μ-almost every x, it satisfies $(\text{aInc})_1$. □

Chapter 3
Generalized Orlicz Spaces

In the previous chapter, we investigated properties of Φ-functions. In this chapter, we use them to derive results for function spaces defined by means of Φ-functions. We first define the spaces and prove the they are quasi-Banach spaces (Sects. 3.1–3.3). Related to the conjugate Φ-functions, we study associate spaces in Sect. 3.4. Separability and uniform convexity require some restrictions on the Φ-function; they are considered in Sects. 3.5 and 3.6. Finally, in Sect. 3.7 we introduce the first of our central conditions, (A0), and prove density of smooth functions.

Recall that we always assume that (A, Σ, μ) is a σ-finite, complete measure space and that μ is not identically zero.

3.1 Modulars

Modular spaces and modular function spaces have been investigated by many researchers in an abstract setting, starting from appropriate axioms, see, e.g., the monographs [20, 71, 74, 96, 98, 99]. We do not treat the abstract case, but rather focus on the particular modular defined by a Φ-function:

Definition 3.1.1 Let $\varphi \in \Phi_{\mathrm{w}}(A, \mu)$ and let ϱ_φ be given by

$$\varrho_\varphi(f) := \int_A \varphi(x, |f(x)|) \, d\mu(x)$$

for all $f \in L^0(A, \mu)$. The function ϱ_φ is called a *modular*. The set

$$L^\varphi(A, \mu) := \left\{ f \in L^0(A, \mu) \, : \, \varrho_\varphi(\lambda f) < \infty \text{ for some } \lambda > 0 \right\}$$

© Springer Nature Switzerland AG 2019
P. Harjulehto, P. Hästö, *Orlicz Spaces and Generalized Orlicz Spaces*,
Lecture Notes in Mathematics 2236,
https://doi.org/10.1007/978-3-030-15100-3_3

is called a *generalized Orlicz space*. If the set and measure are obvious from the context we abbreviate $L^\varphi(A, \mu) = L^\varphi$.

Note that many sources require more conditions for ϱ_φ to be called a modular: φ should be convex and $\varrho_\varphi(f) = 0$ should imply $f = 0$. These conditions hold if $\varphi \in \Phi_c(A, \mu)$ is strictly increasing. Some might call the function ϱ_φ from the previous definition a quasisemimodular, but we prefer the simpler name "modular". The modular ϱ is said to be *left-continuous* if $\lim_{\lambda \to 1^-} \varrho(\lambda f) = \varrho(f)$.

Generalized Orlicz spaces are also called *Musielak–Orlicz spaces*. They provide a good framework for many function spaces.

Example 3.1.2 Let us consider the Φ-functions from Example 2.1.4, namely

$$\varphi^p(t) := \frac{1}{p} t^p, \quad p \in (0, \infty),$$

$$\varphi_{\max}(t) := (\max\{0, t - 1\})^2,$$

$$\varphi_{\sin}(t) := t + \sin(t),$$

$$\varphi_{\exp}(t) := e^t - 1,$$

$$\varphi^\infty(t) := \infty \cdot \chi_{(1,\infty)}(t),$$

$$\varphi^{\infty,2}(t) := \varphi^\infty(t) + \frac{2t-1}{1-t} \chi_{(\frac{1}{2},1)}(t).$$

These generate the Orlicz spaces

$$L^{\varphi^p} = L^p, \quad L^{\varphi_{\max}} = L^2 + L^\infty, \quad L^{\varphi_{\sin}} = L^1, \quad L^{\varphi_{\exp}} = \exp L$$

$$L^{\varphi^\infty} = L^{\varphi^{\infty,2}} = L^\infty.$$

Note that there is no x-dependence in these examples. Typical generalized Orlicz spaces are the variable exponent Lebesgue space $L^{p(\cdot)}$ given by $\varphi(x, t) = t^{p(x)}$ and the double phase space given by the functional $\varphi(x, t) = t^p + a(x)t^q$, see Example 2.5.3.

Lemma 3.1.3 *Let $\varphi \in \Phi_w(A, \mu)$.*

(a) *Then $L^\varphi(A, \mu) = \left\{ f \in L^0(A, \mu) : \lim_{\lambda \to 0^+} \varrho_\varphi(\lambda f) = 0 \right\}$.*

(b) *If, additionally, φ satisfies (aDec), then*

$$L^\varphi(A, \mu) = \left\{ f \in L^0(A, \mu) : \varrho_\varphi(f) < \infty \right\}.$$

Proof

(a) If $\lim_{\lambda \to 0^+} \varrho_\varphi(\lambda f) = 0$, then there exists $\lambda > 0$ such that $\varrho_\varphi(\lambda f) < \infty$. Hence

$$\{f \in L^0(A, \mu) \; : \; \lim_{\lambda \to 0^+} \varrho_\varphi(\lambda f) = 0\} \subset L^\varphi(A, \mu).$$

For the other direction, suppose there exists $\lambda > 0$ such that $\varrho_\varphi(\lambda f) < \infty$. Then (aInc)$_1$ implies that

$$\varphi(x, t\lambda |f(x)|) \leqslant at\varphi(x, \lambda |f(x)|)$$

for $t \in (0, 1)$ and μ-almost all $x \in A$. This yields that

$$\int_A \varphi(x, t\lambda |f(x)|) \, d\mu(x) \leqslant at \int_A \varphi(x, \lambda |f(x)|) \, d\mu(x)$$

and hence $\lim_{t \to 0^+} \varrho_\varphi(tf) = 0$.

(b) The inclusion $\{f \in L^0(A, \mu) \; : \; \varrho_\varphi(f) < \infty\} \subset L^\varphi(A, \mu)$ is obvious. For the other direction, suppose there exists $\lambda \in (0, 1)$ such that $\varrho_\varphi(\lambda f) < \infty$ (the case $\lambda \geqslant 1$ is clear). Then (aDec)$_q$ implies

$$\varrho_\varphi(f) \leqslant \int_A a\lambda^{-q} \varphi(x, \lambda |f|) \, d\mu(x) = a\lambda^{-q} \varrho_\varphi(\lambda f) < \infty.$$

$\square$

The next lemma collects analogues of the classical Lebesgue integral convergence results. These properties are called *Fatou's lemma*, *monotone convergence* and *dominated convergence* for the modular, respectively.

Lemma 3.1.4 *Let $\varphi \in \Phi_w(A, \mu)$ and $f_k, f, g \in L^0(A, \mu)$. In (a) and (b), we assume also that φ is left-continuous.*

(a) *If $f_k \to f$ μ-almost everywhere, then $\varrho_\varphi(f) \leqslant \liminf_{k \to \infty} \varrho_\varphi(f_k)$.*

(b) *If $|f_k| \nearrow |f|$ μ-almost everywhere, then $\varrho_\varphi(f) = \lim_{k \to \infty} \varrho_\varphi(f_k)$.*

(c) *If $f_k \to f$ μ-almost everywhere, $|f_k| \leqslant |g|$ μ-almost everywhere, and $\varrho_\varphi(\lambda g) < \infty$ for every $\lambda > 0$, then $\lim_{k \to \infty} \varrho_\varphi(\lambda |f - f_k|) = 0$ for every $\lambda > 0$.*

Proof Let us first prove (a). Since φ is left-continuous, the mapping $\varphi(x, \cdot)$ is lower semicontinuous by Lemma 2.1.5. Thus Fatou's lemma implies

$$\varrho_\varphi(f) = \int_A \varphi(x, \lim_{k \to \infty} |f_k|) \, d\mu \leqslant \int_A \liminf_{k \to \infty} \varphi(x, |f_k|) \, d\mu$$

$$\leqslant \liminf_{k \to \infty} \int_A \varphi(x, |f_k|) \, d\mu = \liminf_{k \to \infty} \varrho_\varphi(f_k).$$

To prove (b), let $|f_k| \nearrow |f|$. If $\varrho_\varphi(f) = \infty$, then by (a) the claim holds. So let us assume that $\varrho_\varphi(f) < \infty$. By the left-continuity and monotonicity of $\varphi(x, \cdot)$, we have $0 \leqslant \varphi(x, |f_k|) \nearrow \varphi(x, |f|)$ μ-almost everywhere. So, monotone convergence gives

$$\varrho_\varphi(f) = \int_A \varphi(x, \lim_{k \to \infty} |f_k|) \, d\mu = \int_A \lim_{k \to \infty} \varphi(x, |f_k|) \, d\mu$$

$$= \lim_{k \to \infty} \int_A \varphi(x, |f_k|) \, d\mu = \lim_{k \to \infty} \varrho_\varphi(f_k).$$

To prove (c), assume that $f_k \to f$ μ-almost everywhere, $|f_k| \leqslant |g|$, and $\varrho(\lambda g) < \infty$ for every $\lambda > 0$. Then $|f_k - f| \to 0$ μ-almost everywhere, $|f| \leqslant |g|$ and $\lambda|f_k - f| \leqslant 2\lambda|g|$. Since $\varrho_\varphi(2\lambda g) < \infty$, we can use dominated convergence to conclude that

$$\lim_{k \to \infty} \varrho_\varphi(\lambda|f - f_k|) = \int_A \varphi\left(x, \lim_{k \to \infty} \lambda|f - f_k|\right) d\mu = 0.$$

$\square$

In the next lemma we require (Dec) rather than (aDec). Recall that the latter implies the former if φ is convex (Lemma 2.2.6).

Question 3.1.5 Is the next lemma true if we assume (aDec) instead of (Dec)?

Lemma 3.1.6 *Let $\varphi \in \Phi_w(A, \mu)$ satisfy (Dec). Let $f_j, g_j \in L^\varphi(\mathbb{R}^n)$ for $j = 1, 2, \ldots$ with $(\varrho_\varphi(f_j))_{j=1}^\infty$ bounded. If $\varrho_\varphi(f_j - g_j) \to 0$ as $j \to \infty$, then*

$$|\varrho_\varphi(f_j) - \varrho_\varphi(g_j)| \to 0 \quad as \ j \to \infty.$$

Proof Using that φ is increasing and and satisfies (Dec)$_q$, we obtain

$$\varphi(x, g_j) \leqslant \varphi(x, |g_j - f_j| + |f_j|) \leqslant \varphi\left(x, 2|g_j - f_j|\right) + \varphi(x, 2|f_j|)$$

$$\leqslant 2^q \varphi\left(x, |g_j - f_j|\right) + 2^q \varphi(x, |f_j|)$$

and hence $(\varrho_\varphi(g_j))_{j=1}^\infty$ is bounded. Let us choose $c > 0$ such that $\varrho_\varphi(f_j) \leqslant c$ and $\varrho_\varphi(g_j) \leqslant c$. Let $\lambda > 0$ and note that $|f_j| \leqslant |f_j - g_j| + |g_j|$. If $|f_j - g_j| \leqslant \lambda|g_j|$, then by (Dec)$_q$ we have

$$\varphi(x, |f_j|) \leqslant \varphi(x, (1 + \lambda)|g_j|) \leqslant (1 + \lambda)^q \varphi(x, |g_j|).$$

If, on the other hand, $|f_j - g_j| > \lambda|g_j|$, then we estimate by (Dec)$_q$

$$\varphi(x, |f_j|) \leqslant \varphi(x, (1 + \tfrac{1}{\lambda})|f_j - g_j|) \leqslant (1 + \tfrac{1}{\lambda})^q \varphi(x, |f_j - g_j|)$$

Taking into account both cases and integrating over $x \in A$, we find that

$$\varrho_\varphi(f_j) - \varrho_\varphi(g_j) = \int_{\mathbb{R}^n} \varphi(x, |f_j - g_j + g_j|) - \varphi(x, |g_j|) \, d\mu(x)$$
$$\leqslant (1 + \tfrac{1}{\lambda})^q \varrho_\varphi(f_j - g_j) + ((1+\lambda)^q - 1)\varrho_\varphi(g_j).$$

Swapping f_j and g_j gives a similar inequality, and combining the inequalities gives, we find that that

$$|\varrho_\varphi(f_j) - \varrho_\varphi(g_j)| \leqslant (1 + \tfrac{1}{\lambda})^q \varrho_\varphi(f_j - g_j) + ((1+\lambda)^q - 1)(\varrho_\varphi(f_j) + \varrho_\varphi(g_j)).$$

Let $\varepsilon > 0$ be given. Since $\varrho_\varphi(f_j) + \varrho_\varphi(g_j) \leqslant 2c$, we can choose λ so small that

$$((1+\lambda)^q - 1)(\varrho_\varphi(f_j) + \varrho_\varphi(g_j)) \leqslant \frac{\varepsilon}{2}.$$

We can then choose j_0 so large that

$$(1 + \tfrac{1}{\lambda})^q \varrho_\varphi(f_j - g_j) \leqslant \frac{\varepsilon}{2}$$

when $j \geqslant j_0$ and it follows that $|\varrho_\varphi(f_j) - \varrho_\varphi(g_j)| \leqslant \varepsilon$. □

3.2 Quasinorm and the Unit Ball Property

Next we define our quasinorm.

Definition 3.2.1 Let $\varphi \in \Phi_w(A, \mu)$. For $f \in L^0(A, \mu)$ we denote

$$\|f\|_{L^\varphi(A,\mu)} := \inf\left\{ \lambda > 0 : \varrho_\varphi\left(\frac{f}{\lambda}\right) \leqslant 1 \right\}.$$

We abbreviate $\|f\|_{L^\varphi(A,\mu)} = \|f\|_\varphi$ if the set and measure are clear from the context.

Note that we can write the space L^φ with this functional as follows:

$$L^\varphi(A, \mu) = \{f \in L^0(A, \mu) : \|f\|_{L^\varphi(A,\mu)} < \infty\}.$$

Let us show that $\| \cdot \|_{L^\varphi(A,\mu)}$ is in fact a (quasi)norm. For simplicity we will often omit the prefix "quasi" when there is no danger of confusion. For instance, we write "norm convergence" for convergence with respect to $\| \cdot \|_\varphi$, regardless of whether it is a norm or a quasinorm. As is the case for Lebesgue spaces, we identify functions which coincide μ-almost everywhere, since $\| f \|_\varphi = 0$ only implies that $f = 0$ a.e. In the next lemma, we show that $\| \cdot \|_\varphi$ satisfies the quasi-triangle inequality or triangle inequality when $\varphi \in \Phi_w$ or $\varphi \in \Phi_c$, respectively.

Lemma 3.2.2

(a) *If $\varphi \in \Phi_w(A, \mu)$, then $\| \cdot \|_\varphi$ is a quasinorm.*
(b) *If $\varphi \in \Phi_c(A, \mu)$, then $\| \cdot \|_\varphi$ is a norm.*

Proof Assume first that $\varphi \in \Phi_w(A, \mu)$. If $f = 0$ a.e., then $\| f \|_\varphi = 0$. If $\| f \|_\varphi = 0$, then $\varrho_\varphi(\frac{f}{\lambda}) \leqslant 1$ for all $\lambda > 0$. When $f(x) \neq 0$, we have $\frac{|f(x)|}{\lambda} \to \infty$ when $\lambda \to 0^+$. Since $\lim_{t \to \infty} \varphi(x, t) = \infty$ for μ-almost every x, we obtain that $f(x) = 0$ for μ-almost every $x \in A$.

Let $f \in L^\varphi(A, \mu)$ and $a \in \mathbb{R}$. By definition, $\varrho_\varphi(f) = \varrho_\varphi(|f|)$. With the change of variables $\lambda' := \lambda/|a|$, we have

$$\| af \|_\varphi = \inf\left\{ \lambda > 0 : \varrho_\varphi\Big(\frac{af}{\lambda}\Big) \leqslant 1 \right\} = \inf\left\{ \lambda > 0 : \varrho_\varphi\Big(\frac{f}{\lambda/|a|}\Big) \leqslant 1 \right\}$$

$$= |a| \inf\left\{ \lambda' > 0 : \varrho_\varphi\Big(\frac{f}{\lambda'}\Big) \leqslant 1 \right\} = |a| \| f \|_\varphi.$$

Hence $\| \cdot \|_\varphi$ is homogeneous.

Let $f, g \in L^\varphi(A, \mu)$ and $u > \| f \|_\varphi$ and $v > \| g \|_\varphi$. Then $\varrho_\varphi(f/u) \leqslant 1$ and $\varrho_\varphi(g/v) \leqslant 1$ by the definition of the norm. By (aInc)$_1$,

$$\varphi\Big(x, \frac{|f|}{2au}\Big) \leqslant \frac{1}{2}\varphi\Big(x, \frac{|f|}{u}\Big) \quad \text{and} \quad \varphi\Big(x, \frac{|g|}{2av}\Big) \leqslant \frac{1}{2}\varphi\Big(x, \frac{|g|}{v}\Big).$$

Thus we obtain that

$$\int_A \varphi\Big(x, \frac{|f+g|}{4a(u+v)}\Big) d\mu \leqslant \int_A \varphi\Big(x, \frac{|2f|}{4au}\Big) + \varphi\Big(x, \frac{|2g|}{4av}\Big) d\mu$$

$$\leqslant \frac{1}{2} \int_A \varphi\Big(x, \frac{|f|}{u}\Big) + \varphi\Big(x, \frac{|g|}{v}\Big) d\mu$$

$$\leqslant \frac{1}{2} + \frac{1}{2} = 1$$

and hence $\| f + g \|_\varphi \leqslant 4au + 4av$ which yields that $\| f + g \|_\varphi \leqslant 4a\| f \|_\varphi + 4a\| g \|_\varphi$. This completes the proof of (a).

For (b), assume that φ is convex. Let $u > \|f\|_\varphi$ and $v > \|g\|_\varphi$. By the convexity of φ,

$$\int_A \varphi\left(x, \frac{|f+g|}{u+v}\right) d\mu \leqslant \int_A \varphi\left(x, \frac{u}{u+v}\frac{|f|}{u} + \frac{v}{u+v}\frac{|g|}{v}\right) d\mu$$

$$\leqslant \int_A \frac{u}{u+v}\varphi\left(x, \frac{|f|}{u}\right) + \frac{v}{u+v}\varphi\left(x, \frac{|g|}{v}\right) d\mu$$

$$\leqslant \frac{u}{u+v} + \frac{v}{u+v} = 1.$$

Thus $\|f+g\|_\varphi \leqslant u + v$, which yields $\|f+g\|_\varphi \leqslant \|f\|_\varphi + \|g\|_\varphi$, as required for (b). $\qquad\square$

The following is a fundamental relation between the norm and the modular. When working without left-continuity, we need to pay closer attention to when the norm is strictly smaller than 1.

Lemma 3.2.3 (Unit Ball Property) *Let $\varphi \in \Phi_w(A, \mu)$. Then*

$$\|f\|_\varphi < 1 \quad \Rightarrow \quad \varrho_\varphi(f) \leqslant 1 \quad \Rightarrow \quad \|f\|_\varphi \leqslant 1.$$

If φ is left-continuous, then $\varrho_\varphi(f) \leqslant 1 \Leftrightarrow \|f\|_\varphi \leqslant 1$.

Proof If $\varrho_\varphi(f) \leqslant 1$, then $\|f\|_\varphi \leqslant 1$ by definition of $\|\cdot\|_\varphi$. If, on the other hand, $\|f\|_\varphi < 1$, then $\varrho_\varphi(f/\lambda) \leqslant 1$ for some $\lambda < 1$. Since ϱ is increasing, it follows that $\varrho_\varphi(f) \leqslant 1$.

If $\|f\|_\varphi \leqslant 1$, then $\varrho_\varphi(f/\lambda) \leqslant 1$ for all $\lambda > 1$. When ϱ is left-continuous it follows that $\varrho_\varphi(f) \leqslant 1$. $\qquad\square$

The next example show that $\|f\|_\varphi = 1$ does not imply $\varrho_\varphi(f) \leqslant 1$ if the Φ-function is not left-continuous. Let $\varphi(t) := \infty\chi_{[1,\infty)}(t)$ and $f \equiv 1$. Then $\varphi \in \Phi_w$ and $\varrho_\varphi(f) = \infty$. Since $\varrho_\varphi(f/\lambda) \leqslant 1$ if and only if $\lambda > 1$, we have $\|f\|_\varphi = 1$.

We stated previously that equivalent Φ-functions give rise to the same space. Let us now prove this.

Proposition 3.2.4 *Let $\varphi, \psi \in \Phi_w(A, \mu)$. If $\varphi \simeq \psi$, then $L^\varphi(A, \mu) = L^\psi(A, \mu)$ and the norms are comparable.*

Proof Assume that $\psi(x, \frac{t}{L}) \leqslant \varphi(x, t) \leqslant \psi(x, Lt)$ and $f \in L^\varphi(A, \mu)$. Then there exists $\lambda > 0$ such that $\varrho_\psi(\frac{\lambda}{L} f) \leqslant \varrho_\varphi(\lambda f) < \infty$. Thus $f \in L^\psi(A, \mu)$. The other direction is similar and hence $L^\varphi(A, \mu) = L^\psi(A, \mu)$ as sets.

Let $\varepsilon > 0$ and $\lambda = \|f\|_\varphi + \varepsilon$. Then

$$\varrho_\psi\left(\frac{f}{L\lambda}\right) \leqslant \varrho_\varphi\left(\frac{f}{\lambda}\right) \leqslant 1$$

and hence $\|f\|_\psi \leqslant L\lambda = L(\|f\|_\varphi + \varepsilon)$. Letting $\varepsilon \to 0^+$ we obtain that $\|f\|_\psi \leqslant L\|f\|_\varphi$. The other direction is similar and so the norms are comparable. $\square$

For $\varphi \in \Phi_w(A, \mu) \setminus \Phi_c(A, \mu)$, $\|\cdot\|_\varphi$ is a quasinorm, not a norm, but still it has the following quasi-convexity property.

Corollary 3.2.5 *Let* $\varphi \in \Phi_w(A, \mu)$. *Then*

$$\left\| \sum_{i=1}^\infty f_i \right\|_\varphi \lesssim \sum_{i=1}^\infty \|f_i\|_\varphi.$$

Proof By Theorem 2.5.10, there exists $\psi \in \Phi_s(A, \mu)$ such that $\varphi \simeq \psi$. Thus $\|\cdot\|_\psi$ is a norm by Lemma 3.2.2 and hence $\|\sum_{i=1}^\infty f_i\|_\psi \leqslant \sum_{i=1}^\infty \|f_i\|_\psi$. By Proposition 3.2.4, $\|\cdot\|_\psi$ and $\|\cdot\|_\varphi$ are comparable and hence the claim follows. $\square$

Next we prove a general embedding theorem, refining Proposition 3.2.4. The argument follows [34, Theorem 2.8.1].

Note that the inequality implies the embedding even without the extra atom-less assumption. A measure μ is called *atom-less* if for any measurable set A with $\mu(A) > 0$ there exists a measurable subset A' of A such that $\mu(A) > \mu(A') > 0$.

Theorem 3.2.6 *Let* $\varphi, \psi \in \Phi_w(A, \mu)$ *and let the measure* μ *be atom-less. Then* $L^\varphi(A, \mu) \hookrightarrow L^\psi(A, \mu)$ *if and only if there exist* $K > 0$ *and* $h \in L^1(A, \mu)$ *with* $\|h\|_1 \leqslant 1$ *such that*

$$\psi\left(x, \frac{t}{K}\right) \leqslant \varphi(x, t) + h(x)$$

for μ-*almost all* $x \in A$ *and all* $t \geqslant 0$.

Proof Let us start by showing that the inequality implies the embedding. Let $\|f\|_\varphi < 1$, which yields by the unit ball property (Lemma 3.2.3) that $\varrho_\varphi(f) \leqslant 1$. Then, by $(aInc)_1$ and the assumption,

$$\varrho_\psi\left(\frac{f}{2aK}\right) \leqslant a\frac{1}{2a}\varrho_\psi\left(\frac{f}{K}\right) \leqslant \frac{1}{2}\varrho_\varphi(f) + \frac{1}{2}\int_A h\,dy \leqslant 1.$$

This and the unit ball property yield $\|f/(2aK)\|_\psi \leqslant 1$ for $\|f\|_\varphi < 1$. We finish the proof with *the scaling argument*. If $\|f\|_\varphi \geqslant 1$, then we use the above result for $f/(\|f\|_\varphi + \varepsilon)$ which yields the claim as $\varepsilon \to 0^+$ since

$$\left\| \frac{f}{2aK(\|f\|_\varphi + \varepsilon)} \right\|_\psi < 1 \quad \Leftrightarrow \quad \|f\|_\psi < 2aK(\|f\|_\varphi + \varepsilon).$$

Assume next that the embedding $L^\varphi(A, \mu) \hookrightarrow L^\psi(A, \mu)$ holds with embedding constant c_1. By Proposition 3.2.4 and Theorem 2.5.10, we may assume that φ,

$\psi \in \Phi_s(\Omega)$. For $x \in A$ and $t \geqslant 0$ define

$$
\alpha(x, t) := \begin{cases} \psi(x, \frac{t}{c_1}) - \varphi(x, t) & \text{if } \varphi(x, t) < \infty, \\ 0 & \text{if } \varphi(x, t) = \infty. \end{cases}
$$

Since $\varphi(x, \cdot)$ and $\psi(x, \cdot)$ are left-continuous for μ-almost every $x \in A$, also $\alpha(x, \cdot)$ is left-continuous for μ-almost every $x \in A$. Let (r_k) be a sequence of distinct numbers with $\{r_k : k \in \mathbb{N}\} = \mathbb{Q} \cap [0, \infty)$ and $r_1 = 0$. Then

$$
\psi(x, \tfrac{r_k}{c_1}) \leqslant \varphi(x, r_k) + \alpha(x, r_k)
$$

for all $k \in \mathbb{N}$ and μ-almost all $x \in A$. Define

$$
b_k(x) := \max_{1 \leqslant j \leqslant k} \alpha(x, r_j).
$$

Since $r_1 = 0$ and $\alpha(x, 0) = 0$, we have $b_k \geqslant 0$. Moreover, the functions b_k are measurable and increasing in k. The function $b := \sup_k b_k$ is measurable, non-negative, and satisfies

$$
b(x) = \sup_{t \geqslant 0} \alpha(x, t),
$$

$$
\psi(x, \tfrac{t}{c_1}) \leqslant \varphi(x, t) + b(x)
$$

for μ-almost all $x \in A$ and all $t \geqslant 0$, where we have used the left-continuity of $\alpha(x, \cdot)$ and the density of $\{r_k : k \in \mathbb{N}\}$ in $[0, \infty)$.

We now show that $b \in L^1(A, \mu)$ with $\|b\|_1 \leqslant 1$. We consider first the case $|b| < \infty$ μ-almost everywhere, and assume to the contrary that there exists $\varepsilon > 0$ such that

$$
\int_A b \, d\mu \geqslant 1 + 2\varepsilon.
$$

Define

$$
V_k := \{x \in A : \alpha(x, r_k) > \tfrac{1}{1+\varepsilon} b(x)\},
$$

$$
W_{k+1} := V_{k+1} \setminus (V_1 \cup \cdots \cup V_k)
$$

for all $k \in \mathbb{N}$. Note that $V_1 = \emptyset$ due to the special choice $r_1 = 0$. Since $\{r_k : k \in \mathbb{N}\}$ is dense in $[0, \infty)$ and $\alpha(x, \cdot)$ is left-continuous for every $x \in A$, we have $\bigcup_{k=1}^{\infty} V_k = \bigcup_{k=2}^{\infty} W_k = \{x \in A : b(x) > 0\}$.

Let $f := \sum_{k=2}^{\infty} r_k \chi_{W_k}$. For every $x \in W_k$ we have $\alpha(x, r_k) > 0$ and therefore $\varphi(x, r_k) < \infty$. If x is not in $\bigcup_{k=2}^{\infty} W_k$, then $\varphi(x, |f(x)|) = 0$. This implies that $\varphi(x, |f(x)|)$ is everywhere finite. Moreover, by the definition of W_k and α we get

$$\psi\left(x, \frac{|f(x)|}{c_1}\right) \geqslant \varphi(x, |f(x)|) + \frac{1}{1+\varepsilon} b(x) \tag{3.2.1}$$

for μ-almost all $x \in A$.

If $\varrho_\varphi(f) \leqslant 1$, then $\varrho_\psi(\frac{f}{c_1}) \leqslant 1$ by the unit ball property since c_1 is the embedding constant. However, this contradicts

$$\varrho_\psi(\tfrac{f}{c_1}) \geqslant \varrho_\varphi(f) + \frac{1}{1+\varepsilon} \int_A b \, d\mu \geqslant \frac{1+2\varepsilon}{1+\varepsilon} > 1,$$

where we have used (3.2.1) and $\bigcup_{k=2}^{\infty} W_k = \{x \in A : b(x) > 0\}$. So we can assume that $\varrho_\varphi(f) > 1$. Since μ is atom-less and $\varphi(x, |f(x)|)$ is μ-almost everywhere finite, there exists $U \subset A$ with $\varrho_\varphi(f \chi_U) = 1$. Thus by (3.2.1)

$$\varrho_\psi(\tfrac{f}{c_1} \chi_U) \geqslant \varrho_\varphi(f \chi_U) + \tfrac{1}{1+\varepsilon} \int_U b \, d\mu$$
$$= 1 + \tfrac{1}{1+\varepsilon} \int_U b \, d\mu. \tag{3.2.2}$$

Now, $\varrho_\varphi(f \chi_U) = 1$ implies that $\mu(U \cap \{f \neq 0\}) > 0$. Since $\{f \neq 0\} = \bigcup_{k=2}^{\infty} W_k = \{b > 0\}$ we get $\mu(U \cap \{b > 0\}) > 0$ and

$$\int_U b \, d\mu > 0.$$

This and (3.2.2) imply that $\varrho_\psi\left(\frac{1}{c_1} f \chi_U\right) > 1$. On the other hand, $L^\varphi \hookrightarrow L^\psi$, $\varrho_\varphi(f \chi_U) = 1$ and the unit ball property yield that $\varrho_\psi\left(\frac{1}{c_1} f \chi_U\right) \leqslant 1$, which is a contradiction. Thus the case where $|b| < \infty$ μ-almost everywhere is complete.

If we assume that there exists $E \subset A$ with $b|_E = \infty$ and $\mu(E) > 0$, then a similar argument with $V_k := \{x \in E : \alpha(x, r_k) \geqslant \frac{2}{\mu(E)}\}$ yields a contradiction. Hence this case cannot occur, and the proof is complete by what was shown previously. $\qquad\square$

Theorem 3.2.6 yields the following corollary.

Corollary 3.2.7 *Let $\varphi, \psi \in \Phi_w(A, \mu)$, $\varphi \sim \psi$. Then $L^\varphi(A, \mu) = L^\psi(A, \mu)$ and the norms are comparable.*

We also have a different line of results following from the unit ball property. When working with estimates based on $(\text{aInc})_1$, as we mostly do, the left-continuity plays no role, as the following result shows.

Corollary 3.2.8 *Let $\varphi \in \Phi_w(A, \mu)$ and $f \in L^\varphi(A, \mu)$ and let a be the constant from* (aInc)$_1$.

(a) *If $\|f\|_\varphi < 1$, then $\varrho_\varphi(f) \leqslant a\|f\|_\varphi$.*
(b) *If $\|f\|_\varphi > 1$, then $\|f\|_\varphi \leqslant a\varrho_\varphi(f)$.*
(c) *In any case, $\|f\|_\varphi \leqslant a\varrho_\varphi(f) + 1$.*

Proof The claim (a) is obvious for $f = 0$, so let us assume that $0 < \|f\|_\varphi < 1$. Let $\lambda > 1$ be so small that $\lambda\|f\|_\varphi < 1$. By unit ball property (Lemma 3.2.3) and $\left\|\frac{f}{\lambda\|f\|_\varphi}\right\|_\varphi < 1$, it follows that $\varrho_\varphi\left(\frac{f}{\lambda\|f\|_\varphi}\right) \leqslant 1$. Since $\lambda\|f\|_\varphi \leqslant 1$, it follows from (aInc)$_1$ that

$$\frac{1}{a\lambda\|f\|_\varphi}\varrho_\varphi(f) \leqslant \varrho_\varphi\left(\frac{f}{\lambda\|f\|_\varphi}\right) \leqslant 1.$$

And hence (a) follows as $\lambda \to 1^+$.

For (b) assume that $\|f\|_\varphi > 1$. Then $\varrho_\varphi(\frac{f}{\lambda}) > 1$ for $1 < \lambda < \|f\|_\varphi$ and by (aInc)$_1$ it follows that

$$\frac{a}{\lambda}\varrho_\varphi(f) \geqslant \varrho_\varphi\left(\frac{f}{\lambda}\right) > 1.$$

As $\lambda \to \|f\|_\varphi^-$ we obtain that $a\varrho_\varphi(f) \geqslant \|f\|_\varphi$.

Claim (c) follows immediately from (b). □

Note that in Corollary 3.2.8(a) and (b) the case $\|f\|_\varphi = 1$ is excluded. Indeed, let $f \equiv 1$, $\varphi(x, t) := \infty\chi_{(1,\infty)}(t)$ and $\psi(x, t) := \infty\chi_{[1,\infty)}(t)$. Then φ and ψ are Φ-functions and $\|f\|_\varphi = \|f\|_\psi = 1$ but $\varrho_\varphi(f) = 0$ and $\varrho_\psi(f) = \infty$.

Let us next investigate the relationship between norm and modular further, refining Corollary 3.2.8.

Lemma 3.2.9 *Let $\varphi \in \Phi_w(A, \mu)$ satisfy* (aInc)$_p$ *and* (aDec)$_q$, $1 \leqslant p \leqslant q < \infty$. *Then*

$$\min\left\{(\tfrac{1}{a}\varrho_\varphi(f))^{\frac{1}{p}}, (\tfrac{1}{a}\varrho_\varphi(f))^{\frac{1}{q}}\right\} \leqslant \|f\|_\varphi \leqslant \max\left\{(a\varrho_\varphi(f))^{\frac{1}{p}}, (a\varrho_\varphi(f))^{\frac{1}{q}}\right\}$$

for $f \in L^0(A, \mu)$, where a is the maximum of the constants from (aInc)$_p$ *and* (aDec)$_q$.

Proof We start with the second inequality. Let $u > \varrho_\varphi(f)$ and assume first that $au \leqslant 1$. Then (aDec)$_q$ gives that

$$\varphi\left(x, \frac{|f(x)|}{(au)^{\frac{1}{q}}}\right) \leqslant \frac{a}{au}\varphi(x, |f(x)|) = \frac{1}{u}\varphi(x, |f(x)|).$$

Integrating over A, we find that $\varrho_\varphi\big(f/(au)^{1/q}\big) \leqslant 1$, which yields $\|f\|_\varphi \leqslant (au)^{\frac{1}{q}}$. If $au > 1$, we similarly use (aInc)$_p$ to conclude that $\|f\|_\varphi \leqslant (au)^{\frac{1}{p}}$. The second inequality follows as $u \to \varrho_\varphi(f)^+$.

Let us then prove the first inequality. Let $u \in (0, \varrho_\varphi(f))$ and assume first that $\frac{u}{a} \leqslant 1$. Then (aInc)$_p$ gives that

$$\varphi\left(x, \frac{|f(x)|}{(u/a)^{\frac{1}{p}}}\right) \geqslant \frac{a}{au}\varphi(x, |f(x)|) = \frac{1}{u}\varphi(x, |f(x)|).$$

Integrating over A, we find that $\varrho_\varphi\big(f/(u/a)^{1/p}\big) > 1$, which yields $\|f\|_\varphi \geqslant (u/a)^{\frac{1}{p}}$. If $\frac{u}{a} > 1$, we similarly use (aDec)$_q$ to conclude that $\|f\|_\varphi \geqslant (\frac{u}{a})^{\frac{1}{q}}$. The first inequality follows as $u \to \varrho_\varphi(f)^-$.   □

When $q = \infty$, we get the following corollary.

Corollary 3.2.10 *Let* $\varphi \in \Phi_{\mathrm{w}}(A, \mu)$ *satisfy* (aInc)$_p$, $1 \leqslant p < \infty$. *Then*

$$\min\left\{(\tfrac{1}{a}\varrho_\varphi(f))^{\frac{1}{p}}, 1\right\} \leqslant \|f\|_\varphi \leqslant \max\left\{(a\varrho_\varphi(f))^{\frac{1}{p}}, 1\right\}$$

for $f \in L^0(A, \mu)$, *where* a *is the constant from* (aInc)$_p$.

The following result is the generalization of the classical Hölder inequality $\int |f||g|\,d\mu \leqslant \|f\|_p\|g\|_{p'}$ to generalized Orlicz spaces.

Lemma 3.2.11 (Hölder's Inequality) *Let* $\varphi \in \Phi_{\mathrm{w}}(A, \mu)$. *Then*

$$\int_A |f||g|\,d\mu \leqslant 2\|f\|_\varphi \|g\|_{\varphi^*}$$

for all $f \in L^\varphi(A, \mu)$ *and* $g \in L^{\varphi^*}(A, \mu)$. *Moreover, the constant* 2 *cannot in general be replaced by any smaller number.*

Proof Let $f \in L^\varphi$ and $g \in L^{\varphi^*}$ with $u > \|f\|_\varphi$ and $v > \|g\|_{\varphi^*}$. By the unit ball property, $\varrho_\varphi(f/u) \leqslant 1$ and $\varrho_{\varphi^*}(g/v) \leqslant 1$. Thus, using Young's inequality (2.4.1), we obtain

$$\int_A \frac{|f|}{u}\frac{|g|}{v}\,d\mu \leqslant \int_A \varphi\left(x, \frac{|f|}{u}\right) + \varphi^*\left(x, \frac{|g|}{v}\right)d\mu = \varrho_\varphi\left(\frac{f}{u}\right) + \varrho_{\varphi^*}\left(\frac{g}{v}\right) \leqslant 2.$$

Multiplying by uv, we get the inequality as $u \to \|f\|_\varphi^+$ and $v \to \|g\|_{\varphi^*}^+$.

The next example shows that the extra constant 2 in Hölder's inequality cannot be omitted. Let $\varphi(t) = \frac{1}{2}t^2$. Then a short calculation gives that $\varphi^*(t) = \sup_{u \geq 0}(ut - \frac{1}{2}u^2) = \frac{1}{2}t^2$. Let $f \equiv g \equiv 1$. Then $\int_0^1 fg\,dy = 1$. On the other hand,

$$\inf\left\{\lambda > 0 : \int_0^1 \frac{1}{2}\left(\frac{1}{\lambda}\right)^2 dy \leq 1\right\} = \frac{1}{\sqrt{2}}$$

and thus $\|f\|_{L^\varphi(0,1)} = \|g\|_{L^{\varphi^*}(0,1)} = \frac{1}{\sqrt{2}}$ and $\|f\|_{L^\varphi(0,1)}\|g\|_{L^{\varphi^*}(0,1)} = \frac{1}{2}$. $\square$

3.3 Convergence and Completeness

Norm-convergence can be expressed using only the modular, as the following lemma shows. This is often useful, since calculating the precise value of the norm may be difficult.

Lemma 3.3.1 *Let* $\varphi \in \Phi_w(A, \mu)$. *Then* $\|f_k\|_\varphi \to 0$ *as* $k \to \infty$ *if and only if* $\lim_{k \to \infty} \varrho_\varphi(\lambda f_k) = 0$ *for all* $\lambda > 0$.

Proof Assume that $\|f_k\|_\varphi \to 0$. Let $K > 1$ and $\lambda > 0$. Then $\|K\lambda f_k\|_\varphi < 1$ for large k. Thus $\varrho_\varphi(K\lambda f_k) \leq 1$ for large k by unit ball property (Lemma 3.2.3). Hence by (aInc)$_1$

$$\varrho_\varphi(\lambda f_k) = \int_A \varphi(x, \lambda|f_k|)\,d\mu \leq \int_A \frac{a}{K}\varphi(x, K|\lambda f_k|)\,d\mu$$

$$= \frac{a}{K}\varrho_\varphi(K\lambda f_k) \leq \frac{a}{K}$$

for all $K > 1$ and all large k. This implies $\varrho_\varphi(\lambda f_k) \to 0$.

Assume now that $\varrho_\varphi(\lambda f_k) \to 0$ for all $\lambda > 0$. Then $\varrho_\varphi(\lambda f_k) \leq 1$ for large k. By unit ball property (Lemma 3.2.3), $\|f_k\|_\varphi \leq 1/\lambda$ for the same k. Since $\lambda > 0$ was arbitrary, we get $\|f_k\|_\varphi \to 0$. $\square$

Definition 3.3.2 *Let* $\varphi \in \Phi_w(A, \mu)$ *and* $f_k, f \in L^\varphi(A)$. *We say that* f_k *is* modular convergent *(ϱ_φ-convergent) to* f *if* $\varrho(\lambda(f_k - f)) \to 0$ *as* $k \to \infty$ *for some* $\lambda > 0$.

It is clear from Lemma 3.3.1 that modular convergence is weaker than norm convergence. Indeed, for norm convergence we have $\lim_{k \to \infty} \varrho(\lambda(x_k - x)) = 0$ for all $\lambda > 0$, while for modular convergence this only has to hold for some $\lambda > 0$.

In some cases, modular convergence and norm convergence coincide and in others they differ:

Lemma 3.3.3 *Let* $\varphi \in \Phi_w(A, \mu)$. *Modular convergence and norm convergence are equivalent if and only if* $\varrho_\varphi(f_k) \to 0$ *implies* $\varrho_\varphi(2f_k) \to 0$.

Proof Let modular convergence and norm convergence be equivalent and let $\varrho(f_k) \to 0$ with $f_k \in L^\varphi$. Then $f_k \to 0$ (norm convergence) and by Lemma 3.3.1 it follows that $\varrho(2f_k) \to 0$.

Let us then assume that $\varrho_\varphi(f_k) \to 0$ implies $\varrho_\varphi(2f_k) \to 0$. Let $f_k \in L^\varphi$ with $\varrho_\varphi(\lambda_0 f_k) \to 0$ for some $\lambda_0 > 0$. We have to show that $\varrho_\varphi(\lambda f_k) \to 0$ for all $\lambda > 0$. For fixed $\lambda > 0$ choose $m \in \mathbb{N}$ such that $2^m \lambda_0 \geqslant \lambda$. Then by repeated application of the assumption we get $\lim_{k\to\infty} \varrho(2^m \lambda_0 f_k) = 0$. Since φ is increasing we obtain $0 \leqslant \lim_{k\to\infty} \varrho_\varphi(\lambda f_k) \leqslant \lim_{k\to\infty} \varrho_\varphi(2^m \lambda_0 f_k) = 0$. This proves that $f_k \to 0$ by Lemma 3.3.1.   □

Lemmas 2.2.6 and 3.3.3 yield the following corollary.

Corollary 3.3.4 *Let $\varphi \in \Phi_w$ satisfy* (aDec). *Then modular convergence and norm convergence are equivalent.*

Note that if φ satisfies only the weak doubling condition Δ_2^w from Definition 2.5.22, then the conclusion of Corollary 3.3.4 does not hold. This can be seen from the example $\varphi(x, t) := (t - h(x))_+$, where $h \in L^1$ is positive. Now if $f_k = h$, then $\varrho_\varphi(f_k) = 0$ but $\varrho_\varphi(2f_k) = \|h\|_1 > 0$.

Our proof that L^φ is a (quasi-)Banach space essentially follows [34, Theorem 2.3.13]. We start with convergence in measure.

Lemma 3.3.5 *Let $\varphi \in \Phi_w(A, \mu)$ and $\mu(A) < \infty$. Then every $\|\cdot\|_\varphi$-Cauchy sequence is also a Cauchy sequence with respect to convergence in measure.*

Proof Fix $\varepsilon > 0$ and let $V_t := \{x \in A : \varphi(x, t) = 0\}$ for $t > 0$. Then V_t is measurable. For μ-almost all $x \in A$ the function $t \mapsto \varphi(x, t)$ is increasing, so $V_t \subset V_s \cup F$ for all $t > s$ with $\mu(F) = 0$ and F independent of s and t. Since $\lim_{t\to\infty} \varphi(x, t) = \infty$ for μ-almost every $x \in A$ and $\mu(A) < \infty$, we obtain that $\lim_{k\to\infty} \mu(V_k) = 0$. Thus, there exists $K \in \mathbb{N}$ such that $\mu(V_K) < \varepsilon$.

For a μ-measurable set $E \subset A$ define

$$\nu_K(E) := \varrho_\varphi(K\chi_E) = \int_E \varphi(x, K)\, d\mu.$$

If E is μ-measurable with $\nu_K(E) = 0$, then $\varphi(x, K) = 0$ for μ-almost every $x \in E$. Thus $\mu(E \setminus V_K) = 0$ by the definition of V_K. Hence, E is a $\mu|_{A\setminus V_K}$-null set, which means that the measure $\mu|_{A\setminus V_K}$ is absolutely continuous with respect to ν_K.

Since $\mu(A \setminus V_K) \leqslant \mu(A) < \infty$ and $\mu|_{A\setminus V_K}$ is absolutely continuous with respect to ν_K, there exists $\delta \in (0, 1)$ such that $\nu_K(E) \leqslant \delta$ implies $\mu(E \setminus V_K) \leqslant \varepsilon$ (e.g. [50, Theorem 30.B]). Since f_k is a $\|\cdot\|_\varphi$-Cauchy sequence, there exists $k_0 \in \mathbb{N}$ such that $\|\frac{Ka}{\varepsilon\delta}(f_m - f_k)\|_\varphi < 1$ for all $m, k \geqslant k_0$, with a from (aInc)$_1$. Assume in the following that $m, k \geqslant k_0$. Then (aInc)$_1$ and the unit ball property (Lemma 3.2.3) imply

$$\varrho_\varphi\Big(\frac{K}{\varepsilon}(f_m - f_k)\Big) \leqslant \delta\varrho_\varphi\Big(\frac{Ka}{\varepsilon\delta}(f_m - f_k)\Big) \leqslant \delta.$$

Let us write $E_{m,k,\varepsilon} := \{x \in A : |f_m(x) - f_k(x)| \geq \varepsilon\}$. Then

$$\nu_K(E_{m,k,\varepsilon}) = \int_{E_{m,k,\varepsilon}} \varphi(x, K) \, d\mu(x) \leq \varrho_\varphi\left(\tfrac{K}{\varepsilon}(f_m - f_k)\right) \leq \delta.$$

By the choice of δ, this implies that $\mu(E_{m,k,\varepsilon} \setminus V_K) \leq \varepsilon$. With $\mu(V_K) < \varepsilon$ we have $\mu(E_{m,k,\varepsilon}) \leq 2\varepsilon$. Since $\varepsilon > 0$ was arbitrary, this proves that f_k is a Cauchy sequence with respect to convergence in measure. □

Lemma 3.3.6 *Let $\varphi \in \Phi_w(A, \mu)$. Then every $\|\cdot\|_\varphi$-Cauchy sequence $(f_k) \subset L^\varphi$ has a subsequence which converges μ-a.e. to a measurable function f.*

Proof Recall that μ is σ-finite. Let $A := \bigcup_{i=1}^\infty A_i$ with A_i pairwise disjoint and $\mu(A_i) < \infty$ for all $i \in \mathbb{N}$. Then, by Lemma 3.3.5, (f_k) is a Cauchy sequence with respect to convergence in measure on A_1. Therefore there exists a measurable function $f : A_1 \to \mathbb{R}$ and a subsequence of (f_k) which converges to f μ-almost everywhere. Repeating this argument for every A_i and passing to the diagonal sequence we get a subsequence (f_{k_j}) and a μ-measurable function $f : A \to \mathbb{R}$ such that $f_{k_j} \to f$ μ-almost everywhere. □

Let us now prove the completeness of L^φ.

Theorem 3.3.7

(a) *If $\varphi \in \Phi_w(A, \mu)$, then $L^\varphi(A, \mu)$ is a quasi-Banach space.*
(b) *If $\varphi \in \Phi_c(A, \mu)$, then $L^\varphi(A, \mu)$ is a Banach space.*

Proof By Lemma 3.2.2, $\|\cdot\|_\varphi$ is a quasinorm if $\varphi \in \Phi_w(A, \mu)$ and a norm if $\varphi \in \Phi_c(A, \mu)$. It remains to prove completeness.

Let (f_k) be a Cauchy sequence. By Lemma 3.3.6, there exists a subsequence f_{k_j} and a μ-measurable function $f : A \to \mathbb{R}$ such that $f_{k_j} \to f$ for μ-almost every $x \in A$. This implies $\varphi(x, c|f_{k_j}(x) - f(x)|) \to 0$ μ-almost everywhere for every $c > 0$. Let $\lambda > 0$ and $0 < \varepsilon < 1$. Since (f_k) is a Cauchy sequence, there exists $N = N(\lambda, \varepsilon) \in \mathbb{N}$ such that $\|\lambda(f_m - f_k)\|_\varphi < \varepsilon/a$ for all $m, k \geq N$, with a from (aInc)$_1$. By Corollary 3.2.8(a) this implies $\varrho_\varphi(\lambda(f_m - f_k)) \leq \varepsilon$ for all $m, k \geq N$. Since φ is increasing, we obtain $\varphi(x, \lim_{j\to\infty} \tfrac{\lambda}{2}|f_m - f_{k_j}|) \leq \liminf_{j\to\infty} \varphi(x, \lambda|f_m - f_{k_j}|)$. Hence Fatou's lemma yields that

$$\varrho_\varphi\left(\tfrac{\lambda}{2}(f_m - f)\right) = \int_A \varphi\left(x, \lim_{j\to\infty} \tfrac{\lambda}{2}|f_m - f_{k_j}|\right) d\mu$$

$$\leq \int_A \liminf_{j\to\infty} \varphi(x, \lambda|f_m - f_{k_j}|) \, d\mu$$

$$\leq \liminf_{j\to\infty} \int_A \varphi(x, \lambda|f_m - f_{k_j}|) \, d\mu \leq \varepsilon.$$

Thus $\varrho_\varphi(\tfrac{\lambda}{2}(f_m - f)) \to 0$ for $m \to \infty$ and every $\lambda > 0$, so that $\|f_m - f\|_\varphi \to 0$ by Lemma 3.3.1. Therefore every Cauchy sequence converges in L^φ. □

Let us summarize a few additional properties of L^φ. Let $\varphi \in \Phi_w(A, \mu)$. Then $L^\varphi(A, \mu)$ is *circular*, i.e.

$$\|f\|_\varphi = \big\| |f| \big\|_\varphi \quad \text{for all} \quad f \in L^\varphi. \tag{3.3.1}$$

If $f \in L^\varphi$, $g \in L^0(A, \mu)$, and $0 \leqslant |g| \leqslant |f|$ μ-almost everywhere, then Then $L^\varphi(A, \mu)$ is *solid*, i.e.

$$g \in L^\varphi \quad \text{and} \quad \|g\|_\varphi \leqslant \|f\|_\varphi. \tag{3.3.2}$$

Next lemma gives more properties of $\| \cdot \|_\varphi$, namely *Fatou's lemma (for the (quasi)norm)* and the *Fatou property*. The corresponding result for the modular was proved in Lemma 3.1.4.

Lemma 3.3.8 *Let $\varphi \in \Phi_w(A, \mu)$ be left-continuous and $f, f_k \in L^0(A, \mu)$.*

(a) *If $f_k \to f$ μ-almost everywhere, then $\|f\|_\varphi \leqslant \liminf_{k\to\infty} \|f_k\|_\varphi$.*
(b) *If $|f_k| \nearrow |f|$ μ-almost everywhere with $f_k \in L^\varphi(A, \mu)$ and $\sup_k \|f_k\|_\varphi < \infty$, then $f \in L^\varphi(A, \mu)$ and $\|f_k\|_\varphi \nearrow \|f\|_\varphi$.*

Proof For (a) let $f_k \to f$ μ-almost everywhere. There is nothing to prove for $\liminf_{k\to\infty} \|f_k\|_\varphi = \infty$. Otherwise, let $\lambda > \liminf_{k\to\infty} \|f_k\|_\varphi$. Then $\|f_k\|_\varphi < \lambda$ for some large k. Thus by the unit ball property (Lemma 3.2.3), $\varrho_\varphi(f_k/\lambda) \leqslant 1$ for large k. Now Fatou's lemma for the modular (Lemma 3.1.4) implies $\varrho_\varphi(f/\lambda) \leqslant 1$. So $\|f\|_\varphi \leqslant \lambda$ again by the unit ball property. Thus we have $\|f\|_\varphi \leqslant \liminf_{k\to\infty} \|f_k\|_\varphi$.

It remains to prove (b). So let $|f_k| \nearrow |f|$ μ-almost everywhere with $\sup_k \|f_k\|_\varphi < \infty$. By (a) we obtain $\|f\|_\varphi \leqslant \liminf_{k\to\infty} \|f_k\|_\varphi \leqslant \sup_k \|f_k\|_\varphi < \infty$, which also proves $f \in L^\varphi$. On the other hand, $|f_k| \nearrow |f|$ and solidity (3.3.2) implies that $\|f_k\|_\varphi \nearrow \limsup_{k\to\infty} \|f_k\|_\varphi \leqslant \|f\|_\varphi$. It follows that $\lim_{k\to\infty} \|f_k\|_\varphi = \|f\|_\varphi$ and $\|f_k\|_\varphi \nearrow \|f\|_\varphi$. $\square$

3.4 Associate Spaces

The dual space can be used to understand properties of a vector space. We define the related concept of associate space, which is more closely connected to the scale of generalized Orlicz spaces. In particular, we will show that the second associate space is always isomorphic to the space itself, whereas the second dual space is only isomorphic under certain additional conditions.

The dual space X^* of a normed space X consists of all bounded linear functions from X to $\mathbb{R}$. Equipped with the norm

$$\|f\|_{X^*} := \sup_{\|f\|_X \leqslant 1} |F(f)|,$$

X^* is a Banach space, see for example [122, Theorem 4.1, p. 92].

Definition 3.4.1 Let $\varphi \in \Phi_{\mathrm{w}}(A, \mu)$. Then by $(L^\varphi(A, \mu))^*$ we denote the dual space of $L^\varphi(A, \mu)$. Furthermore, we define $\sigma_\varphi : (L^\varphi(A, \mu))^* \to [0, \infty]$ by

$$\sigma_\varphi(F) := \sup_{f \in L^\varphi(A,\mu)} \big(|F(f)| - \varrho_\varphi(f)\big).$$

Note the difference between the spaces $(L^\varphi(A, \mu))^*$ and $L^{\varphi^*}(A, \mu)$: the former is the dual space of $L^\varphi(A, \mu)$, whereas the latter is the generalized Orlicz space defined by the conjugate modular φ^*.

By definition of the functional σ_φ we have

$$|F(f)| \leqslant \varrho_\varphi(f) + \sigma_\varphi(F) \tag{3.4.1}$$

for all $f \in L^\varphi(A, \mu)$ and $F \in (L^\varphi(A, \mu))^*$. This is a generalized version of the classical Young inequality.

Remark 3.4.2 The function σ_φ is actually a semimodular on the dual space. We refer to [34, Section 2.7] for details.

In the definition of σ_φ the supremum is taken over all $L^\varphi(A, \mu)$. However, it is possible to restrict this to the closed unit ball when F is in the unit ball and φ is convex.

Lemma 3.4.3 *Let $\varphi \in \Phi_{\mathrm{c}}(A, \mu)$. If $F \in (L^\varphi(A, \mu))^*$ with $\|F\|_{(L^\varphi)^*} \leqslant 1$, then*

$$\sigma_\varphi(F) = \sup_{f \in L^\varphi, \|f\|_\varphi \leqslant 1} \big(|F(f)| - \varrho(f)\big) = \sup_{f \in L^\varphi, \varrho_\varphi(f) \leqslant 1} \big(|F(f)| - \varrho(f)\big).$$

Proof The equivalence of the suprema follows from the unit ball property (Lemma 3.2.3). Let $\|F\|_{(L^\varphi)^*} \leqslant 1$. By the definition of the dual norm we have

$$\sup_{\|f\|_\varphi > 1} \big(|F(f)| - \varrho(f)\big) \leqslant \sup_{\|f\|_\varphi > 1} \big(\|F\|_{(L^\varphi)^*} \|f\|_\varphi - \varrho_\varphi(f)\big)$$

$$\leqslant \sup_{\|f\|_\varphi > 1} \big(\|f\|_\varphi - \varrho_\varphi(f)\big).$$

If $\|f\|_\varphi > 1$, then $\varrho_\varphi(f) \geqslant \|f\|_\varphi$ by Corollary 3.2.8, and so the right-hand side of the previous inequality is non-positive. Since ϱ^* is defined as a supremum, and is always non-negative, we see that f with $\|f\|_\varphi > 1$ does not affect the supremum, and so the claim follows.   □

The next lemma shows that we can approximate the function 1 with a monotonically increasing sequence of functions in the generalized Orlicz space. This will allow us to generalize several results from [34, Chapter 2] without the extraneous assumption $L^\infty \subset L^\varphi$ that was used there.

Lemma 3.4.4 *Let $\varphi \in \Phi_w(A, \mu)$. There exist a sequence of positive functions $h_k \in L^\varphi(A, \mu)$, $k \in \mathbb{N}$, such that $h_k \nearrow 1$ and $\{h_k = 1\} \nearrow A$.*

Proof We set $h(x) := \varphi^{-1}(x, 1)$. Then h is measurable by Lemma 2.5.12 and $\varphi(x, h(x)) \leqslant 1$ by Lemma 2.3.9(b). Let us define $h_k := \min\{kh\chi_{B(0,k)\cap A}, 1\}$. Then

$$\varrho_\varphi(\tfrac{1}{k}h_k) \leqslant \int_{B(0,k)\cap A} \varphi(x, \min\{h, 1/k\})\,dx \leqslant |B(0,k)| < \infty,$$

so that $h_k \in L^\varphi(A)$. By $\lim_{t\to 0^+} \varphi(x, t) = 0$ we have $h > 0$. It follows that $kh\chi_{B(0,k)\cap A} \nearrow \infty$ for μ-almost every $x \in A$, and so $\{h_k = 1\} \nearrow A \setminus F$, $\mu(F) = 0$. By modifying h_k in a set of measure zero, we obtain the claim. $\qquad\square$

Definition 3.4.5 We define the *associate space* of $L^\varphi(A, \mu)$ as the space $(L^\varphi)'(A, \mu) := \{f \in L^0(A, \mu) : \|f\|_{(L^\varphi)'} < \infty\}$ with the norm

$$\|f\|_{(L^\varphi)'} := \sup_{\|g\|_\varphi \leqslant 1} \int_A fg\,d\mu.$$

If $g \in (L^\varphi)'$ and $f \in L^\varphi$, then $fg \in L^1$ by the definition of the associate space. In particular, the integral $\int_A fg\,d\mu$ is well defined and

$$\left| \int_A fg\,d\mu \right| \leqslant \|g\|_{(L^\varphi)'}\|f\|_{L^\varphi}.$$

By J_f we denote the functional $g \mapsto \int_A fg\,d\mu$. Clearly $J_f \in (L^\varphi)^*$ when $f \in (L^\varphi)'$ so $J. : (L^\varphi)' \to (L^\varphi)^*$. The next result shows that the associate space of L^φ is always given by L^{φ^*}. In this sense the associate space is much nicer than the general dual space.

Theorem 3.4.6 (Norm Conjugate Formula) *If $\varphi \in \Phi_w(A, \mu)$, then $(L^\varphi)' = L^{\varphi^*}$ and the norms are comparable. Moreover, for all $f \in L^0(A, \mu)$*

$$\|f\|_\varphi \approx \sup_{\|g\|_{\varphi^*} \leqslant 1} \int_A |fg|\,d\mu.$$

Proof By Theorem 2.5.10 there exists $\psi \in \Phi_s(A, \mu)$ such that $\varphi \simeq \psi$. Then $L^\varphi = L^\psi$ and $\|f\|_\varphi \approx \|f\|_\psi$ by Proposition 3.2.4.

Let $f \in (L^\psi)'$ with $\|f\|_{(L^\psi)'} \leqslant 1$ and $\varepsilon > 0$. Let $\{q_1, q_2, \dots\}$ be an enumeration of non-negative rational numbers with $q_1 = 0$. For $k \in \mathbb{N}$ and $x \in A$ define

$$r_k(x) := \max_{j\in\{1,\dots,k\}} \{q_j |f(x)| - \psi(x, q_j)\}.$$

The special choice $q_1 = 0$ implies that $r_k(x) \geqslant 0$ for all $x \geqslant 0$. Since $\mathbb{Q}$ is dense in $[0, \infty)$ and $\psi(x, \cdot)$ is left-continuous, $r_k(x) \nearrow \psi^*(x, |f(x)|)$ for μ-almost every $x \in A$ as $k \to \infty$.

Since f and $\psi(\cdot, t)$ are measurable functions, the sets

$$E_{i,k} := \left\{ x \in A \: : \: q_i \, |f(x)| - \psi(x, q_i) = \max_{j=1,\ldots,k} \left(q_j \, |f(x)| - \psi(x, q_j) \right) \right\}$$

are measurable. Let $F_{i,k} := E_{i,k} \setminus (E_{1,k} \cup \ldots \cup E_{i-1,k})$ and define

$$g_k := \sum_{i=1}^{k} q_i \chi_{F_{i,k}}.$$

Then g_k is measurable and bounded and

$$r_k(x) = g_k(x) \, |f(x)| - \psi(x, g_k(x))$$

for all $x \in A$.

Let $h_k \in L^\psi(A, \mu)$ be as in Lemma 3.4.4. Since g_k is bounded and $h_k \in L^\psi(A, \mu)$, it follows that $w := \operatorname{sgn}(f) \, h_k g_k \in L^\psi(A, \mu)$.

Since σ_ψ is defined in Definition 3.4.1 as a supremum over functions in L^ψ, we get a lower bound by using the particular function $w \chi_E$. Thus

$$\sigma_\psi(J_f) \geqslant |J_f(w \chi_{\{h_k=1\}})| - \varrho_\psi(w \chi_{\{h_k=1\}}) = \int_{\{h_k=1\}} fw - \psi(x, |w|) \, d\mu$$

$$\geqslant \int_{\{h_k=1\}} g_k \, |f| - \psi(x, |g_k|) \, d\mu = \int_A r_k \chi_{\{h_k=1\}} \, d\mu.$$

Since $r_k \chi_{\{h_k=1\}} \nearrow \psi^*(x, |f|)$ μ-almost everywhere, it follows by monotone convergence that $\sigma_\psi(J_f) \geqslant \varrho_{\psi^*}(f)$. From the definitions of σ_ψ and ϱ_{ψ^*} we conclude by Young's inequality (2.4.1) that

$$\sigma_\psi(J_f) = \sup_{g \in L^\psi} \int_A fg - \psi(x, g) \, d\mu \leqslant \sup_{g \in L^\psi} \int_A \psi^*(x, f) \, d\mu = \varrho_{\psi^*}(f).$$

Hence $\sigma_\psi(J_f) = \varrho_{\psi^*}(f)$.

Recall that we are assuming $\|f\|_{(L^\psi)'} \leqslant 1$ and denote $G := \{ g \in L^\psi \: : \: \|g\|_\psi \leqslant 1 \}$. Then Lemma 3.4.3 and the definition of the associate space yield

$$\sigma_\psi(J_f) = \sup_{g \in G} \left(|J_f(g)| - \varrho_\psi(g) \right) \leqslant \sup_{g \in G} \left(\|g\|_\psi - \varrho_\psi(g) \right) \leqslant \sup_{g \in G} \|g\|_\psi \leqslant 1.$$

Hence also $\varrho_{\psi^*}(f) = \sigma_\psi(J_f) \leqslant 1$ and it follows from the unit-ball property that $\|f\|_{\psi^*} \leqslant 1$. By a scaling argument, we obtain $\|f\|_{\psi^*} \leqslant \|f\|_{(L^\psi)'}$.

Hölder's inequality (Lemma 3.2.11) implies that $\|f\|_{(L^\psi)'} \leqslant 2\|f\|_{\psi^*}$. In view of the previous paragraph, $\|f\|_{(L^\psi)'} \approx \|f\|_{\psi^*}$

Taking into account that $\psi^{**} \simeq \psi$ (Proposition 2.4.5), we have shown that $L^\varphi = L^\psi = (L^{\psi^*})'$. By the definition of the associate space norm, this means that

$$\|f\|_\varphi \approx \|f\|_\psi \approx \sup_{\|g\|_{\psi^*} \leqslant 1} \int_A |f|\,|g|\,d\mu$$

for $f \in L^\psi$. By Lemma 2.4.3, $\psi^* \simeq \varphi^*$ and hence $\|g\|_{\psi^*} \approx \|g\|_{\varphi^*}$ (Proposition 3.2.4). By $\frac{1}{L}\|g\|_{\psi^*} \leqslant \|g\|_{\varphi^*} \leqslant L\|g\|_{\psi^*}$ we obtain

$$\sup_{\|g\|_{\varphi^*} \leqslant 1} \int_A |f|\,|g|\,d\mu \geqslant \sup_{L\|g\|_{\psi^*} \leqslant 1} \int_A |f|\,|g|\,d\mu = \frac{1}{L} \sup_{\|Lg\|_{\psi^*} \leqslant 1} \int_A |f|\,|Lg|\,d\mu$$

and similarly for the other direction. Thus the claim is proved in the case $f \in L^\varphi$.

In the case $f \in L^0 \setminus L^\psi$, we can approximate $h_k \min\{|f|, k\} \nearrow |f|$ as before. Since $h_k \min\{|f|, k\} \in L^\psi$, the previous result implies that the formula holds, in the form $\infty = \infty$, when $f \in L^0 \setminus L^\psi$. □

3.5   Separability

In this section we study separability and other density results. Recall that a (quasi-)Banach space is *separable* if it contains a dense, countable subset.

We say that a function is *simple* if it is a linear combination of characteristic functions of measurable sets, $\sum_{i=1}^k s_i \chi_{E_i}(x)$ with $\mu(E_1), \dots, \mu(E_k) < \infty$ and $s_1, \dots, s_k \in \mathbb{R}$. We denote the set of simple functions by $S(A, \mu)$, or, when A and μ are clear, by S. Note that our definition of simple functions includes an assumption that the measures of sets are finite. In many places simple functions are defined without this assumption. Note that a simple function does not necessary belong to L^φ, see Example 3.7.11.

Proposition 3.5.1 *Let $\varphi \in \Phi_w(A, \mu)$ satisfy the assumption* (aDec). *Then the sets $S(A, \mu) \cap L^\varphi(A, \mu)$ and $L^\infty(A, \mu) \cap L^\varphi(A, \mu)$ are dense in $L^\varphi(A, \mu)$.*

Proof Let $f \in L^\varphi(A, \mu)$ with $f \geqslant 0$. Since f is measurable, there exist $f_k := \sum_{i=1}^k s_i \chi_{E_i}(x)$ with measurable sets E_i and $0 \leqslant f_k \nearrow f$ μ-almost everywhere. Note that it does not necessary hold that $\mu(E_i) < \infty$. Since μ is σ-finite, there exist sets (A_i) such that $A = \bigcup_{i=1}^\infty A_i$ and $\mu(A_i) < \infty$ for every i. We define $\tilde{f}_k := \sum_{i=1}^k s_i \chi_{E_i}(x)\chi_{\bigcup_{j=1}^k A_j}(x)$. Then $\tilde{f}_k \in S$ and $0 \leqslant \tilde{f}_k \nearrow f$ μ-almost everywhere. Since $0 \leqslant \tilde{f}_k \leqslant f$ we find that $\tilde{f}_k \in L^\varphi(A, \mu)$. Since φ satisfies (aDec), norm and modular convergence are equivalent by Corollary 3.3.4. Let $\lambda > 0$ be such that $\varrho_\varphi(\lambda f) < \infty$. Then $\lambda|f - \tilde{f}_k| \leqslant \lambda|f|$ and hence by dominated convergence

$\varrho_\varphi(\lambda|f - \tilde{f}_k|) \to 0$ as $k \to \infty$. Since norm and modular convergence are equivalent this yields that $\tilde{f}_k \to f$ in $L^\varphi(A, \mu)$. Thus, f is in the closure of $S \cap L^\varphi(A, \mu)$. If we drop the assumption $f \geqslant 0$, then we obtain the same result by considering the positive and negative parts of f separately.

Since every simple function is bounded, it follows that the larger set $L^\infty \cap L^\varphi$ is also dense in L^φ. □

We say that a measure μ is *separable* if there exists a sequence $(E_k) \subset \Sigma$ with the following properties:

(a) $\mu(E_k) < \infty$ for all $k \in \mathbb{N}$,
(b) for every $E \in \Sigma$ with $\mu(E) < \infty$ and every $\varepsilon > 0$ there exists an index k such that $\mu(E \triangle E_k) < \varepsilon$, where $\triangle$ denotes the symmetric difference defined as $E \triangle E_k := (E \setminus E_k) \cup (E_k \setminus E)$.

For instance the Lebesgue measure on $\mathbb{R}^n$ and the counting measure on $\mathbb{Z}^n$ are separable. Under (aDec), the separability of the measure implies separability of the space. Since L^∞ is not separable, the assumption (aDec) is reasonable.

Theorem 3.5.2 *Let* $\varphi \in \Phi_w(A, \mu)$ *satisfy* (aDec), *and let* μ *be separable. Then* $L^\varphi(A, \mu)$ *is separable.*

Proof Let S_0 be the set of all simple functions of the form $\sum_{i=1}^k a_i \chi_{E_i}$ with $a_i \in \mathbb{Q}$ and E_i is as in the definition of a separable measure, so that S_0 is countable.

By Proposition 3.5.1 it suffices to prove that S_0 is dense in S. Let $f \in S \cap L^\varphi$ be non-negative. Then we can write f in the form $f = \sum_{i=1}^k b_i \chi_{B_i}$ with $b_i \in (0, \infty)$, $B_i \in \Sigma$ pairwise disjoint and $\mu(B_i) < \infty$ for all i. Let h_k be as in Lemma 3.4.4.

Fix $\varepsilon \in (0, 1)$. Let $\lambda \in (0, 1]$ be such that $\varrho_\varphi(\lambda f) < \infty$. By (aDec)$_q$ we obtain

$$\varrho_\varphi(6f\chi_F) = \int_F \varphi(x, 6f)\, d\mu \leqslant \frac{a6^q}{\lambda^q} \int_F \varphi(x, \lambda f)\, d\mu$$

and similarly $\varrho_\varphi(f) < \infty$. By the absolute continuity of the integral we may choose $\delta_1 > 0$ such that

$$\varrho_\varphi(6f\chi_F) < \varepsilon$$

for every measurable set F with $\mu(F) < k\delta_1$. Next choose $l \in \mathbb{N}$ such that $\mu(\bigcup B_i \setminus \{h_l = 1\}) < \frac{1}{2}\delta_1$. By (aDec) and absolute continuity of the integral, we can choose $\delta_2 > 0$ such that

$$\varrho_\varphi(6bh_l\chi_F) < \varepsilon$$

for every measurable set F with $\mu(F) < k\delta_2$, where $b := \max\{b_i\}$. Then choose rational numbers $a_1, \ldots, a_k \in (0, \infty)$ such that $|b_i - a_i| < \varepsilon b_i$ for $i = 1, \ldots, k$. Furthermore, for each i we find j_i such that $\mu(B_i \triangle E_{j_i}) < \min\{\frac{1}{2}\delta_1, \delta_2\}$. Let $g :=$

$h_l \sum_{i=1}^k a_i \chi_{E_{j_i}}$. Then

$$
|f - g| = \left| \sum_{i=1}^k (b_i - a_i) \chi_{B_i} \right| + \left| \sum_{i=1}^k a_i \left(\chi_{B_i} - h_l \chi_{E_{j_i}} \right) \right|
$$

$$
\leqslant \sum_{i=1}^k |b_i - a_i| \chi_{B_i} + \sum_{i=1}^k \left(a_i \, \chi_{B_i \setminus (E_{j_i} \cap \{h_l = 1\})} + h_l a_i \, \chi_{E_{j_i} \setminus B_i} \right)
$$

$$
\leqslant \varepsilon f + 2 \sum_{i=1}^k \left(b_i \, \chi_{B_i \setminus (E_{j_i} \cap \{h_l = 1\})} + b h_l \chi_{E_{j_i} \setminus B_i} \right).
$$

Denote $F := \bigcup_i B_i \setminus (E_{j_i} \cap \{h_l = 1\})$ and $F' := \bigcup_i E_{j_i} \setminus B_i$. Then $\mu(F) \leqslant \sum_{i=1}^k (\mu(B_i \setminus E_{j_i}) + \mu(B_i \setminus \{h_l = 1\})) \leqslant \frac{k}{2}\delta_1 + \frac{k}{2}\delta_1 = k\delta_1$ and $\mu(F') \leqslant k\delta_2$. Taking φ of both sides of the previous estimate for $|f - g|$, and integrating over A, we find by (aDec) that

$$
\varrho_\varphi(f - g) \leqslant \varrho_\varphi(\varepsilon f + 2 f \chi_F + 2 b h_l \chi_{F'})
$$

$$
\leqslant \varrho_\varphi(3\varepsilon f) + \varrho_\varphi(6 f \chi_F) + \varrho_\varphi(6 b h_l \chi_{F'})
$$

$$
\lesssim \varepsilon \varrho_\varphi(f) + 2\varepsilon.
$$

It follows that $\varrho_\varphi(f - g) \to 0$ as $\varepsilon \to 0^+$. Since norm and modular convergence are equivalent (Corollary 3.3.4), this implies the claim. $\square$

3.6 Uniform Convexity and Reflexivity

In this section we prove the reflexivity of L^φ by means of uniform convexity, since it is well known that the latter implies the former. The section is based on [53].

Note that there is no reason to work with non-convex Φ-functions in this context, since mid-point convexity implies convexity for increasing functions. A uniformly convex Φ-function need not to be left-continuous: for example $t \mapsto \infty \chi_{[1,\infty)}(t)$ is uniformly convex. In this section we nevertheless work with Φ_c, for simplicity.

Definition 3.6.1 We say that $\varphi \in \Phi_c(A, \mu)$ is *uniformly convex* if for every $\varepsilon > 0$ there exists $\delta \in (0, 1)$ such that

$$
\varphi\left(x, \frac{s + t}{2}\right) \leqslant (1 - \delta) \frac{\varphi(x, s) + \varphi(x, t)}{2}
$$

for μ-almost every $x \in A$ whenever $s, t \geqslant 0$ and $|s - t| \geqslant \varepsilon \max\{|s|, |t|\}$.

Uniformly convex Φ-functions can be very neatly described in terms of equivalent Φ-functions and (aInc).

Proposition 3.6.2 *The function $\varphi \in \Phi_{\mathrm{w}}(A, \mu)$ is equivalent to a uniformly convex Φ-function if and only it satisfies* (aInc).

Proof Assume first that φ satisfies (aInc)$_p$ with $p > 1$. By Lemma 2.5.9, there exists $\psi \in \Phi_{\mathrm{c}}(A, \mu)$ such that $\varphi \simeq \psi$ and $\psi^{\frac{1}{p}}$ is convex. The claim follows once we show that ψ is uniformly convex. Let $\varepsilon \in (0, 1)$ and $s - t \geqslant \varepsilon s$, with $s > t > 0$. Since $\psi^{\frac{1}{p}}$ is convex,

$$\psi\left(x, \frac{s+t}{2}\right)^{\frac{1}{p}} \leqslant \frac{\psi(x, s)^{\frac{1}{p}} + \psi(x, t)^{\frac{1}{p}}}{2}.$$

Since $t \leqslant (1 - \varepsilon)s$ and ψ is convex, we find that $\psi(x, t) \leqslant \psi(x, (1 - \varepsilon)s) \leqslant (1 - \varepsilon)\psi(x, s)$. Therefore, $\psi(x, t)^{\frac{1}{p}} \leqslant (1 - \varepsilon')\psi(x, s)^{\frac{1}{p}}$ for some $\varepsilon' > 0$ depending only on ε and p. Since $t \mapsto t^p$ is uniformly convex, we obtain that

$$\left(\frac{\psi(x, s)^{\frac{1}{p}} + \psi(x, t)^{\frac{1}{p}}}{2}\right)^p \leqslant (1 - \delta)\frac{\psi(x, s) + \psi(x, t)}{2}.$$

Combined with the previous estimate, this shows that ψ is uniformly convex.

Assume now conversely, that $\varphi \simeq \psi$ and ψ is uniformly convex. Choose $\varepsilon = \frac{1}{2}$ and $t = 0$ in the definition of uniform convexity:

$$\psi(x, \tfrac{s}{2}) \leqslant \tfrac{1}{2}(1 - \delta)\psi(x, s).$$

Divide this equation with $(s/2)^p$ where $p > 1$ is given by $2^{p-1}(1 - \delta) = 1$:

$$\frac{\psi(x, \frac{s}{2})}{(s/2)^p} \leqslant 2^{p-1}(1 - \delta)\frac{\psi(x, s)}{s^p} = \frac{\psi(x, s)}{s^p}.$$

The previous inequality holds for every $s > 0$. If $0 < t < s$, then we can choose $k \in \mathbb{N}$ such that $2^k t \leqslant s < 2^{k+1}t$. Then by the previous inequality and monotonicity of ψ,

$$\frac{\psi(x, t)}{t^p} \leqslant \frac{\psi(x, 2t)}{(2t)^p} \leqslant \cdots \leqslant \frac{\psi(x, 2^k t)}{(2^k t)^p} \leqslant 2^p \frac{\psi(x, s)}{s^p}.$$

Hence ψ satisfies (aInc)$_p$ with $p > 1$. Since this property is invariant under equivalence (Lemma 2.1.8), it holds for φ as well.   □

Of course we are ultimately interested in the uniform convexity of the space. We recall the definition.

Definition 3.6.3 A vector space X is *uniformly convex* if it has a norm $\| \cdot \|$ such that for every $\varepsilon > 0$ there exists $\delta > 0$ with

$$\|x - y\| \geqslant \varepsilon \quad \text{or} \quad \|x + y\| \leqslant 2(1 - \delta)$$

for all $x, y \in X$ with $\|x\| = \|y\| = 1$.

In the Orlicz case, it is well known that the space L^φ is reflexive and uniformly convex if and only if φ and φ^* are doubling [117, Theorem 2, p. 297]. Hudzik [64] showed in 1983 that the same conditions are sufficient for uniform convexity in the generalized Orlicz spaces (see also [41, 65]). With the equivalence technique, we are able to give a simple proof of this result.

The next technical lemma allows us to have absolute values in the inequality from the definition of uniform convexity.

Lemma 3.6.4 *Let $\varphi \in \Phi_c(A, \mu)$ be uniformly convex. Then for every $\varepsilon > 0$ there exists $\delta_2 > 0$ such that*

$$\varphi\left(x, \left|\frac{s+t}{2}\right|\right) \leqslant (1 - \delta_2) \frac{\varphi(x, |s|) + \varphi(x, |t|)}{2}$$

for all $s, t \in \mathbb{R}$ with $|s - t| > \varepsilon \max\{|s|, |t|\}$ and every $x \in A$.

Proof Fix $\varepsilon \in (0, 1)$ and let $\delta > 0$ be as in Definition 3.6.1. Let $|s - t| > \varepsilon \max\{|s|, |t|\}$. If $\left||s| - |t|\right| > \varepsilon \max\{|s|, |t|\}$, then the claim follows by uniform convexity of φ, $|s + t| \leqslant |s| + |t|$ and the choice $\delta_2 := \delta$. So assume in the following $\left||s| - |t|\right| \leqslant \varepsilon \max\{|s|, |t|\}$. Since $|s - t| > \varepsilon \max\{|s|, |t|\}$, it follows that s and t have opposite signs, and that

$$\left|\frac{s+t}{2}\right| = \left|\frac{|s| - |t|}{2}\right| \leqslant \frac{\varepsilon}{2} \max\{|s|, |t|\}.$$

Then it follows from convexity that

$$\varphi\left(x, \left|\frac{s+t}{2}\right|\right) \leqslant \frac{\varepsilon}{2}\varphi(x, \max\{|s|, |t|\}) \leqslant \varepsilon \frac{\varphi(x, |s|) + \varphi(x, |t|)}{2}.$$

Therefore the claim holds with $\delta_2 := \min\{\delta, 1 - \varepsilon\}$.   □

Lemma 3.6.5 *Let $\varphi \in \Phi_c(A, \mu)$ be uniformly convex. Then for every $\varepsilon > 0$ there exists $\delta > 0$ such that*

$$\varrho_\varphi\left(\frac{f - g}{2}\right) < \varepsilon \frac{\varrho_\varphi(f) + \varrho_\varphi(g)}{2} \quad \text{or} \quad \varrho_\varphi\left(\frac{f + g}{2}\right) \leqslant (1 - \delta)\frac{\varrho_\varphi(f) + \varrho_\varphi(g)}{2}$$

for all $f, g \in L^0(A, \mu)$.

Proof Fix $\varepsilon > 0$. Let $\delta_2 > 0$ be as in Lemma 3.6.4 for $\varepsilon/4$. There is nothing to show if $\varrho_\varphi(f) = \infty$ or $\varrho_\varphi(g) = \infty$. So in the following let $\varrho_\varphi(f), \varrho_\varphi(g) < \infty$, which imply by convexity that $\varrho_\varphi(\frac{f+g}{2}), \varrho_\varphi(\frac{f-g}{2}) < \infty$.

Assume that $\varrho_\varphi(\frac{f-g}{2}) \geq \varepsilon \frac{\varrho_\varphi(f)+\varrho_\varphi(g)}{2}$. We show that the second inequality in the statement of the lemma holds with $\delta = \frac{\delta_2 \varepsilon}{2}$. Define

$$E := \left\{ x \in A : |f(x) - g(x)| > \tfrac{\varepsilon}{2} \max\{|f(x)|, |g(x)|\} \right\}.$$

By $(\mathrm{Inc})_1$, for μ-almost all $x \in A \setminus E$, we have

$$\varphi\left(x, \frac{|f(x) - g(x)|}{2}\right) \leq \frac{\varepsilon}{4} \varphi(x, \max\{|f(x)|, |g(x)|\})$$

$$\leq \frac{\varepsilon}{2} \frac{\varphi(x, |f(x)|) + \varphi(x, |g(x)|)}{2}.$$

It follows that

$$\varrho_\varphi\left(\chi_{A \setminus E} \frac{f-g}{2}\right) \leq \frac{\varepsilon}{2} \frac{\varrho_\varphi(\chi_{A \setminus E} f) + \varrho_\varphi(\chi_{A \setminus E} g)}{2} \leq \frac{\varepsilon}{2} \frac{\varrho_\varphi(f) + \varrho_\varphi(g)}{2}.$$

This and $\varrho_\varphi(\frac{f-g}{2}) \geq \varepsilon \frac{\varrho_\varphi(f)+\varrho_\varphi(g)}{2}$ imply

$$\varrho_\varphi\left(\chi_E \frac{f-g}{2}\right) = \varrho_\varphi\left(\frac{f-g}{2}\right) - \varrho_\varphi\left(\chi_{A \setminus E} \frac{f-g}{2}\right) \geq \frac{\varepsilon}{2} \frac{\varrho_\varphi(f) + \varrho_\varphi(g)}{2}. \tag{3.6.1}$$

On the other hand it follows by the definition of E and the choice of δ_2 in Lemma 3.6.4 that

$$\varrho_\varphi\left(\chi_E \frac{f+g}{2}\right) \leq (1 - \delta_2) \frac{\varrho_\varphi(\chi_E f) + \varrho_\varphi(\chi_E g)}{2}. \tag{3.6.2}$$

Since $\frac{1}{2}(\varphi(x, f) + \varphi(x, g)) - \varphi(x, \frac{f+g}{2}) \geq 0$ on $A \setminus E$ (by convexity), we obtain

$$\frac{\varrho_\varphi(f) + \varrho_\varphi(g)}{2} - \varrho_\varphi\left(\frac{f+g}{2}\right) \geq \frac{\varrho_\varphi(\chi_E f) + \varrho_\varphi(\chi_E g)}{2} - \varrho_\varphi\left(\chi_E \frac{f+g}{2}\right).$$

This, (3.6.2), convexity and (3.6.1) imply

$$\frac{\varrho_\varphi(f) + \varrho_\varphi(g)}{2} - \varrho_\varphi\left(\frac{f+g}{2}\right) \geq \delta_2 \frac{\varrho_\varphi(\chi_E f) + \varrho_\varphi(\chi_E g)}{2}$$

$$\geq \delta_2 \varrho_\varphi\left(\chi_E \frac{f-g}{2}\right) \geq \frac{\delta_2 \varepsilon}{2} \frac{\varrho_\varphi(f) + \varrho_\varphi(g)}{2}.$$

$\square$

The question arises whether uniform convexity of φ implies the uniform convexity of L^φ. This turns out to be true under the (aDec) condition. Note that compared to [34, Corollary 2.7.18] here we do not assume that simple functions belong to $L^\varphi(A, \mu)$ and to its dual.

Theorem 3.6.6 *Let $\varphi \in \Phi_c(A, \mu)$ be uniformly convex and satisfy* (aDec). *Then $L^\varphi(A, \mu)$ is uniformly convex with norm $\|\cdot\|_\varphi$.*

In particular, if φ satisfies (aInc) *and* (aDec), *then $L^\varphi(A, \mu)$ is uniformly convex and reflexive.*

Proof Fix $\varepsilon > 0$. Let $f, g \in L^\varphi(A, \mu)$ with $\|f\|_\varphi, \|g\|_\varphi \leqslant 1$ and $\|f - g\|_\varphi > \varepsilon$. Then $\|\frac{f-g}{2}\|_\varphi > \frac{\varepsilon}{2}$ and by Lemma 3.2.9 there exists $\alpha = \alpha(\varepsilon) > 0$ such that $\varrho_\varphi(\frac{f-g}{2}) > \alpha$. By the unit ball property (Lemma 3.2.3) we have $\varrho_\varphi(f), \varrho_\varphi(g) \leqslant 1$, so $\varrho_\varphi(\frac{f-g}{2}) > \alpha \frac{\varrho_\varphi(f)+\varrho_\varphi(g)}{2}$. By Lemma 3.6.5, there exists $\beta = \beta(\alpha) > 0$ such that $\varrho_\varphi(\frac{f+g}{2}) \leqslant (1 - \beta)\frac{\varrho_\varphi(f)+\varrho_\varphi(g)}{2} \leqslant 1 - \beta$. Since φ is a convex Φ-function, it satisfies (Inc) and by Lemma 2.2.6 (aDec) implies (Dec). Now Lemma 3.2.9 implies the existence of $\delta = \delta(\beta) > 0$ with $\|\frac{f+g}{2}\|_\varphi \leqslant 1 - \delta$. This proves the uniform convexity of the norm $\|\cdot\|_\varphi$.

If φ satisfies (aInc) and (aDec), then it is equivalent to some $\psi \in \Phi_c(A, \mu)$ which is uniformly convex and satisfies (aDec), by Proposition 3.6.2. Hence by the first part L^ψ is uniformly convex and by Proposition 1.3.5 it is reflexive. Since $L^\varphi = L^\psi$ by Proposition 3.2.4, the same holds for L^φ.   □

The conditions (aInc) and (aDec) can be generalized further.

Corollary 3.6.7 *Let $\varphi \in \Phi_w(A, \mu)$. If φ satisfies Δ_2^w and ∇_2^w, then $L^\varphi(A, \mu)$ is uniformly convex and reflexive.*

Proof By Theorem 2.5.24, Lemma 2.2.6 and Corollary 2.4.11, there exists $\psi \in \Phi_w(A, \mu)$ which satisfies (aDec), (aInc) and $\varphi \sim \psi$. Hence by Theorem 3.6.6, L^ψ is uniformly convex and reflexive. Since $\varphi \sim \psi$, Corollary 3.2.7 and Proposition 3.2.4 imply that $L^\varphi = L^\psi$, and hence we have proved that L^φ is uniformly convex and reflexive.   □

3.7 The Weight Condition (A0) and Density of Smooth Functions

In this section we introduce a new assumption on the Φ-function and study its implication. The assumption means that we restrict our attention to the essentially "unweighted" case: if $\varphi(x, t) = t^p w(x)$, then (A0) holds if and only if $w \approx 1$.

Definition 3.7.1 We say that $\varphi \in \Phi_w(A, \mu)$ satisfies (A0), if there exists a constant $\beta \in (0, 1]$ such that $\beta \leqslant \varphi^{-1}(x, 1) \leqslant \frac{1}{\beta}$ for μ-almost every $x \in A$.

Equivalently, this means that there exists $\beta \in (0, 1]$ such that $\varphi(x, \beta) \leqslant 1 \leqslant \varphi(x, 1/\beta)$ for μ-almost every $x \in A$ (cf. Corollary 3.7.4).

Example 3.7.2 Let $\varphi(x, t) = \frac{1}{p(x)}t^{p(x)}$ where $p : A \to [1, \infty)$ is measurable, and $\psi(x, t) = t^p + a(x)t^q$ where $1 \leqslant p < q < \infty$ and $a : A \to [0, \infty)$ is measurable. Then $\varphi, \psi \in \Phi_s(A, \mu)$. Since $\varphi^{-1}(x, t) = (p(x)t)^{1/p(x)}$, we see that φ satisfies (A0) (without assumptions for p). By Corollary 3.7.4, ψ satisfies (A0) if and only if $a \in L^\infty(A, \mu)$.

By Theorem 2.3.6 we have $\varphi \simeq \psi$ if and only if $\varphi^{-1} \approx \psi^{-1}$ and thus (A0) is invariant under equivalence of weak Φ-functions.

Note that if φ satisfies (A0), then it is not necessary that $\varphi(x, 1) \approx 1$. For instance, for $\varphi_\infty(t) = \infty\chi_{(1,\infty)}$, we have $\varphi_\infty^{-1}(x, 1) = 1$ whereas φ_∞ only takes values 0 and ∞. However there exists an equivalent weak Φ-function for which also $\varphi(x, 1)$ is controlled.

Lemma 3.7.3 *Let $\varphi \in \Phi_w(A, \mu)$ satisfy (A0). Then there exists $\psi \in \Phi_s(A, \mu)$ with $\varphi \simeq \psi$ and $\psi(x, 1) = \psi^{-1}(x, 1) = 1$ for μ-almost every $x \in A$.*

Proof By Theorem 2.5.10 there exists $\psi_1 \in \Phi_s(A, \mu)$ with $\varphi \simeq \psi_1$. Since φ satisfies (A0) so does ψ_1. We set

$$\psi_2(x, t) := \psi_1(x, \psi_1^{-1}(x, 1)t).$$

By Lemma 2.5.12, $x \mapsto \psi_1^{-1}(x, t)$ is measurable. Thus $x \mapsto \psi_2(x, t)$ is measurable for fixed t by the definition of generalized Φ-prefunction. Then φ_2 satisfies the measurability condition of $\Phi_s(A, \mu)$ by Theorem 2.5.4.

We show that $\psi_2 \in \Phi_s(A, \mu)$. The function ψ_2 is increasing since ψ_1 is increasing. By Lemma 2.5.7, ψ_2 is a Φ-prefunction. Since $t \mapsto \psi_1(x, t)$ is convex we obtain that

$$\psi_2(x, \theta t + (1 - \theta)s) = \psi_1\big(x, \theta\psi_1^{-1}(x, 1)t + (1 - \theta)\psi_1^{-1}(x, 1)s\big)$$
$$\leqslant \theta\psi_1\big(x, \psi_1^{-1}(x, 1)t\big) + (1 - \theta)\psi_1\big(x, \psi_1^{-1}(x, 1)s\big)$$
$$= \theta\psi_2(x, t) + (1 - \theta)\psi_2(x, t)$$

for every $\theta \in [0, 1]$ and $s, t \geqslant 0$. Since $t \mapsto \psi_1(x, t)$ is continuous into the compactification $[0, \infty]$ for μ-almost every x and $\psi_1^{-1}(x, 1)$ is independent of t, we obtain that $t \mapsto \psi_2(x, t)$ is continuous for μ-almost every x.

Since ψ_1 satisfies (A0), we have $\psi_1 \simeq \psi_2$. By Lemma 2.3.3,

$$\psi_2(x, 1) = \psi_1(x, \psi_1^{-1}(x, 1)) = 1$$

for μ-almost every $x \in A$. By Corollary 2.3.4, this implies $\psi_2^{-1}(x, 1) = 1$ for μ-almost every $x \in A$. $\qquad\square$

Corollary 3.7.4 *Let $\varphi \in \Phi_w(A, \mu)$. Then φ satisfies (A0) if and only if there exists $\beta \in (0, 1]$ such that $\varphi(x, \beta) \leqslant 1 \leqslant \varphi(x, 1/\beta)$ for μ-almost every $x \in A$.*

Proof Assume first that (A0) holds. By Lemma 3.7.3, there exists $\psi \in \Phi_s(A, \mu)$ with $\psi(x, 1) = 1$ and $\psi \simeq \varphi$. This implies the inequality.

Assume then that the inequality holds. By the definition of φ^{-1}, the inequality $\varphi(x, \frac{1}{\beta}) \geqslant 1$ yields $\varphi^{-1}(x, 1) \leqslant \frac{1}{\beta}$. By (aInc)$_1$ and $\varphi(x, \beta) \leqslant 1$, we obtain

$$\frac{\varphi(x, \beta/(2a))}{\beta/(2a)} \leqslant a\frac{\varphi(x, \beta)}{\beta} \leqslant \frac{a}{\beta}$$

so that $\varphi(x, \frac{\beta}{2a}) \leqslant \frac{1}{2}$. This yields that $\varphi^{-1}(x, 1) \geqslant \frac{\beta}{2a}$. □

Corollary 3.7.5 *Let* $\varphi \in \Phi_w(A, \mu)$. *If there exists* $c > 0$ *such that* $\varphi(x, c) \approx 1$, *then* φ *satisfies* (A0).

Proof Let $m \leqslant \varphi(x, c) \leqslant M$. We may assume that $m \in (0, 1]$ and $M \geqslant 1$. By (aInc)$_1$ we obtain

$$\frac{\varphi(x, c/(aM))}{c/(aM)} \leqslant a\frac{\varphi(x, c)}{c} \leqslant \frac{aM}{c} \quad \text{and} \quad \frac{m}{c} \leqslant \frac{\varphi(x, c)}{c} \leqslant a\frac{\varphi(x, ac/m)}{ac/m}$$

Thus $\varphi(x, c/(aM)) \leqslant 1$ and $\varphi(x, ac/m) \geqslant 1$ and the claim follows from Corollary 3.7.4. □

Lemma 3.7.6 *If* $\varphi \in \Phi_w(A, \mu)$ *satisfies* (A0), *then* φ^* *satisfies* (A0).

Proof By Theorem 2.4.8 and (A0) of φ we obtain

$$(\varphi^*)^{-1}(x, 1) \approx \frac{1}{\varphi^{-1}(x, 1)} \leqslant \frac{1}{\beta} \quad \text{and} \quad (\varphi^*)^{-1}(x, 1) \approx \frac{1}{\varphi^{-1}(x, 1)} \geqslant \beta$$

for μ-almost every $x \in A$. □

We next characterize the embeddings of the sum and the intersection of generalized Orlicz spaces. Let us introduce the usual notation. Recall that for two normed spaces X and Y (which are both subsets of a vector spaces Z) we equip the *intersection* $X \cap Y$ and the *sum* $X + Y := \{g + h : g \in X, h \in Y\}$ with the norms

$$\|f\|_{X \cap Y} := \max\{\|f\|_X, \|f\|_Y\} \quad \text{and} \quad \|f\|_{X+Y} := \inf_{\substack{f=g+h, \\ g \in X, h \in Y}} (\|g\|_X + \|h\|_Y).$$

In the next lemma we use the convention that every $\varphi \in \Phi_w(A, \mu)$ satisfies (aDec)$_\infty$ with constant 1. Hence we always have $L^1 \cap L^\infty \hookrightarrow L^\varphi \hookrightarrow L^1 + L^\infty$ Notice that the first embedding requires only $\varphi(x, \frac{1}{\beta}) \geqslant 1$ whereas the second requires only $\varphi(x, \beta) \leqslant 1$.

Lemma 3.7.7 *Let* $\varphi \in \Phi_w(A, \mu)$ *satisfy* (A0), (aInc)$_p$ *and* (aDec)$_q$, $p \in [1, \infty)$ *and* $q \in [1, \infty]$. *Then*

$$L^p(A, \mu) \cap L^q(A, \mu) \hookrightarrow L^\varphi(A, \mu) \hookrightarrow L^p(A, \mu) + L^q(A, \mu)$$

and the embedding constants depend only on (A0), (aInc)$_p$ *and* (aDec)$_q$.

Proof Let $\beta \in (0, 1]$ be the constant from Corollary 3.7.4 and a be the maximum of the constant from (aDec) and (aInc).

Let us first study $L^\varphi(A, \mu) \hookrightarrow L^p(A, \mu) + L^q(A, \mu)$. Let $f \in L^\varphi(A, \mu)$ with $\|f\|_\varphi < 1$ so that $\varrho_\varphi(f) \leqslant 1$ by the unit ball property (Lemma 3.2.3). We may assume that $f \geqslant 0$ since otherwise we may study $|f|$. We assume that $p, q \in [1, \infty)$. The cases $q = \infty$ follows by simple modifications.

Define $f_1 := f \chi_{\{0 \leqslant f \leqslant \frac{1}{\beta}\}}$ and $f_2 := f \chi_{\{f > \frac{1}{\beta}\}}$. By Corollary 3.7.4, (aInc)$_p$ and (aDec)$_q$ we have

$$\beta^p \leqslant \frac{\varphi(x, 1/\beta)}{1/\beta^p} \leqslant a \frac{\varphi(x, t)}{t^p} \quad \text{and} \quad \beta^q \leqslant \frac{\varphi(x, 1/\beta)}{1/\beta^q} \leqslant a \frac{\varphi(x, s)}{s^q}$$

for $s \leqslant \frac{1}{\beta} \leqslant t$. Using these we obtain that

$$\frac{\beta^p}{a} \int_A f_1^p \, dx \leqslant \int_A \varphi(x, f_1) \, dx \leqslant 1 \quad \text{and} \quad \frac{\beta^q}{a} \int_A f_2^q \, dx \leqslant \int_A \varphi(x, f_2) \, dx \leqslant 1.$$

Thus we have $\|f\|_{L^p + L^q} \leqslant \frac{a^{1/p}}{\beta} + \frac{a^{1/s}}{\beta}$ and claims follows by the scaling argument, i.e. by using this result for $f/(\|f\|_\varphi + \varepsilon)$ and then letting $\varepsilon \to 0^+$.

Then we consider the embedding $L^p(A, \mu) \cap L^q(A, \mu) \hookrightarrow L^\varphi(A, \mu)$ and assume that $\|f\|_{L^p \cap L^q} \leqslant \frac{1}{a} \min\{\beta^p, \beta^q\}$. Define $f_1 := f \chi_{\{0 \leqslant f \leqslant \beta\}}$ and $f_2 := f \chi_{\{f > \beta\}}$. By Corollary 3.7.4, (aInc)$_p$ and (aDec)$_q$ we have

$$\frac{\varphi(x, t)}{t^p} \leqslant a \frac{\varphi(x, \beta)}{\beta^p} \leqslant \frac{a}{\beta^p} \quad \text{and} \quad \frac{\varphi(x, s)}{s^q} \leqslant a \frac{\varphi(x, \beta)}{\beta^q} \leqslant \frac{a}{\beta^q}$$

for $t \leqslant \beta \leqslant s$. Using these we obtain that

$$\int_A \varphi(x, f_1) \, dx \leqslant \frac{a}{\beta^p} \int_A f_1^p \, dx \leqslant 1 \quad \text{and} \quad \int_A \varphi(x, f_2) \, dx \leqslant \frac{a}{\beta^q} \int_A f_2^q \, dx \leqslant 1.$$

Thus we have $\|f\|_{L^\varphi} \leqslant 1$ and claims follows by the scaling argument. $\quad\square$

Next we give an example which shows that assumption (A0) is not redundant in Lemma 3.7.7.

Example 3.7.8 Let $\varphi(x, t) = t^2 |x|^2$. Then $\varphi \in \Phi_s(\mathbb{R})$ satisfies (Inc)$_2$ and (Dec)$_2$ but not (A0).

First we show that $L^\varphi(\mathbb{R}) \hookrightarrow L^2(\mathbb{R}) + L^\infty(\mathbb{R})$ does not hold. For that let $f(x) := \frac{1}{\sqrt{|x|}} \chi_{(-1,1)}$. Then

$$\int_{\mathbb{R}} \varphi(x, f) \, dx = \int_{(-1,1)} \frac{1}{|x|} |x|^2 \, dx \approx 1$$

and thus $f \in L^\varphi(\mathbb{R})$. Let $f_1 \in L^2(\mathbb{R})$ and $f_2 \in L^\infty(\mathbb{R})$ be such that $f = f_1 + f_2$. Then we find $r > 0$ such that $f(x) = f_1(x)$ for all $x \in (-r, r)$ and obtain

$$\int_{(-r,r)} f_1^2 \, dx = \int_{(-r,r)} \frac{1}{|x|} \, dx = \infty$$

and thus such a decomposition does not exist.

Next we show that $L^1(\mathbb{R}) \cap L^2(\mathbb{R}) \hookrightarrow L^\varphi(\mathbb{R})$ does not hold. Let $g(x) := \min\{1, |x|^{-5/4}\}$. A short calculation shows that $g \in L^1(\mathbb{R})$. Since $0 < g \leqslant 1$ this yields that $g \in L^2(\mathbb{R})$. On the other hand for every $\lambda > 0$ we have

$$\int_\mathbb{R} \varphi(x, \lambda g) \, dx \geqslant \int_{\mathbb{R} \setminus (-1,1)} \left(\lambda |x|^{-\frac{5}{4}}\right)^2 |x|^2 \, dx \approx \lambda^2 \int_1^\infty |x|^{-\frac{1}{2}} \, dx = \infty$$

and thus $g \notin L^\varphi(\mathbb{R})$.

When the set A has finite measure, the previous result simplifies and we get the following corollaries.

Corollary 3.7.9 *Let A have finite measure and let $\varphi \in \Phi_w(A, \mu)$ satisfy (A0) and* $(a\mathrm{Inc})_p$. *Then $L^\varphi(A, \mu) \hookrightarrow L^p(A, \mu)$ and there exists β such that*

$$\int_A |f|^p \, d\mu \lesssim \int_A \varphi(x, |f|) \, d\mu + \mu\left(\{0 < |f| < \tfrac{1}{\beta}\}\right).$$

Proof Let $\beta \in (0, 1]$ be from Corollary 3.7.4. Then by $(a\mathrm{Inc})_p$ and (A0)

$$a \frac{\varphi(x, t)}{t^p} \geqslant \frac{\varphi(x, 1/\beta)}{1/\beta^p} \geqslant \beta^p$$

for all $t \geqslant \frac{1}{\beta}$, so that $a\varphi(x, t) \geqslant \beta^p t^p$. Thus

$$\beta^p t^p \leqslant a\varphi(x, t) + \chi_{\{0 < t < \frac{1}{\beta}\}},$$

which yields the claim for the modulars when we set $t := |f(x)|$ and integrate over $x \in A$. The embedding follows from Lemma 3.7.7 since $L^p(A, \mu) + L^\infty(A, \mu) = L^p(A, \mu)$. □

Similarly, since $L^\infty(A, \mu) = L^p(A, \mu) \cap L^\infty(A, \mu)$ when $\mu(A) < \infty$, Lemma 3.7.7 also implies the following result.

Corollary 3.7.10 *Let A have finite measure and let $\varphi \in \Phi_w(A, \mu)$ satisfy (A0). Then $L^\infty(A, \mu) \hookrightarrow L^\varphi(A, \mu)$.*

The next example shows that the previous result need not hold if φ does not satisfy (A0).

Example 3.7.11 Let $(0, 1) \subset \mathbb{R}$ and $\varphi(x, t) := \frac{t}{|x|}$. Then $\varphi \in \Phi_s(0, 1)$ and φ does not satisfy (A0). Let $f \equiv 2 \in L^\infty(A, \mu)$. We obtain

$$\int_0^1 \varphi(x, \lambda|f|)\, dx = \int_0^1 \frac{2\lambda}{x}\, dx = \infty$$

for all $\lambda > 0$ and hence $f \notin L^\varphi(0, 1)$.

Next we show that $L^\varphi(A, \mu)$ is a Banach function space provided that φ satisfies (A0).

Definition 3.7.12 A normed space $(X, \|\cdot\|_X)$ with $X \subset L^0(A, \mu)$ is called a *Banach function space*, if

(a) $(X, \|\cdot\|_X)$ is circular, solid and satisfies the Fatou property (see p. 62).
(b) If $\mu(E) < \infty$, then $\chi_E \in X$.
(c) If $\mu(E) < \infty$, then $\chi_E \in X'$, i.e. $\int_E |f|\, d\mu \leqslant c(E)\|f\|_X$ for all $f \in X$.

We have seen that the properties in (a) always hold. The next theorem shows that (A0) implies the other two.

Theorem 3.7.13 *Let* $\varphi \in \Phi_w(A, \mu)$ *satisfy* (A0). *Then* $L^\varphi(A, \mu)$ *is a Banach function space.*

Proof Circularity and solidity hold by (3.3.1) and (3.3.2). The Fatou property holds by Lemma 3.3.8. So we only check (b) and (c).

For (b) let $\mu(E) < \infty$. By Corollary 3.7.4 there exists $\beta > 0$ such that $\varphi(x, \beta) \leqslant 1$ and hence

$$\int_A \varphi(x, \beta\chi_E)\, d\mu = \int_E \varphi(x, \beta)\, d\mu \leqslant \mu(E)$$

so that $\chi_E \in L^\varphi(A, \mu)$. By Theorem 3.4.6, $(L^\varphi)' = L^{\varphi^*}$, and by Lemma 3.7.6, φ^* satisfies (A0). Therefore (c) follows from (b) of φ^*. $\square$

At the end of this section we give some basic density results in $\Omega \subset \mathbb{R}^n$ with the Lebesgue measure. Note that the assumption (aDec) is not redundant, since the results do not hold in L^∞. Let us denote by $L_0^\varphi(\Omega)$ the set of functions from $L^\varphi(\Omega)$ whose support is compactly in Ω.

Lemma 3.7.14 *Let* $\varphi \in \Phi_w(\Omega)$ *satisfy* (aDec). *Then* $L_0^\varphi(\Omega)$ *is dense in* $L^\varphi(\Omega)$.

Proof Let $f \in L^\varphi(\Omega)$ and let $\lambda > 0$ be such that $\int_\Omega \varphi(x, \lambda f)\, dx < \infty$. Define $f_i := f\chi_{B(0,i)}$. Then

$$\int_\Omega \varphi(x, \lambda|f - f_i|)\, dx = \int_{\Omega \setminus B(0,i)} \varphi(x, \lambda|f|)\, dx \to 0$$

as $i \to \infty$ by the absolute continuity of the integral. Hence (f_i) is modular convergent to f and thus the claim follows by Corollary 3.3.4.   □

If we also have (A0), then a stronger result holds.

Theorem 3.7.15 *If $\varphi \in \Phi_w(\Omega)$ satisfies (A0) and (aDec), then $C_0^\infty(\Omega)$ is dense in $L^\varphi(\Omega)$.*

Proof Let φ satisfy (aDec)$_q$ and note that simple functions are dense in $L^\varphi(\Omega)$ by Proposition 3.5.1. Since every simple function belongs to $L^1(\Omega) \cap L^q(\Omega)$, it can be approximated by a sequence of $C_0^\infty(\Omega)$ functions in the same space. By Lemma 3.7.7, $L^1(\Omega) \cap L^q(\Omega) \hookrightarrow L^\varphi(\Omega)$ so the claim follows.   □

Chapter 4
Maximal and Averaging Operators

For the rest of the book, we always consider subsets of $\mathbb{R}^n$ and the Lebesgue measure. By Ω we always denote an open set in $\mathbb{R}^n$.

In this chapter we introduce new conditions (A1) and (A2) in Sects. 4.1 and 4.2, which together with (A0) from the previous chapter and (aInc) imply the boundedness of the maximal operator (Sect. 4.3). Additionally, we study averaging operators which are bounded without the (aInc) condition and derive the boundedness of convolution and estimate the norms of characteristic functions (Sect. 4.4).

4.1 The Local Continuity Condition (A1)

We first consider a local condition on the Φ-functions which is relevant when we compare the values of φ at nearby points. It turns out to be convenient to do the comparison for the inverse functions. For some examples of the (A1) condition, see the special cases of variable exponent and double phase growth, collected in Table 7.1 on page 146.

Definition 4.1.1 Let $\varphi \in \Phi_{\mathrm{w}}(\Omega)$. We say that φ satisfies

(A1) if there exists $\beta \in (0, 1)$ such that

$$\beta \varphi^{-1}(x, t) \leqslant \varphi^{-1}(y, t)$$

for every $t \in [1, \frac{1}{|B|}]$, almost every $x, y \in B \cap \Omega$ and every ball B with $|B| \leqslant 1$.

P. Harjulehto, P. Hästö, *Orlicz Spaces and Generalized Orlicz Spaces*,
Lecture Notes in Mathematics 2236,
https://doi.org/10.1007/978-3-030-15100-3_4

(A1$'$) if there exists $\beta \in (0, 1)$ such that

$$\varphi(x, \beta t) \leqslant \varphi(y, t)$$

for every $\varphi(y, t) \in [1, \frac{1}{|B|}]$, almost every $x, y \in B \cap \Omega$ and every ball B with $|B| \leqslant 1$.

One good feature with (A1) is that it is easy to see that it is invariant under equivalence of Φ-functions and works well with conjugation (cf. Lemma 4.1.7). For (A1$'$) we need an additional assumption.

Question 4.1.2 Is (A1$'$) invariant under equivalence of Φ-functions?

Lemma 4.1.3 *Let* $\varphi, \psi \in \Phi_w(\Omega)$ *and* $\varphi \simeq \psi$.

(a) *If* φ *satisfies* (A1)*, then* ψ *does, as well.*
(b) *If* $\varphi \in \Phi_s(\Omega)$ *satisfies* (A1$'$)*, then* ψ *does, as well.*
(c) *If* $\varphi \in \Phi_w(\Omega)$ *satisfies* (A0) *and* (A1$'$)*, then* ψ *does, as well.*

Proof

(a) Suppose that φ satisfies (A1). Since $\varphi \simeq \psi$, Theorem 2.3.6 implies that $\varphi^{-1} \approx \psi^{-1}$. Hence

$$\tfrac{\beta}{L^2}\psi^{-1}(x, t) \leqslant \tfrac{\beta}{L}\varphi^{-1}(x, t) \leqslant \tfrac{1}{L}\varphi^{-1}(y, t) \leqslant \psi^{-1}(y, t)$$

for every $t \in \left[1, \frac{1}{|B|}\right]$, almost every $x, y \in B \cap \Omega$ and every ball B with $|B| \leqslant 1$, and so ψ satisfies (A1).
(b) Suppose then that $\varphi \in \Phi_s(\Omega)$ satisfies (A1$'$) and let t be such that $\psi(y, t) \in [1, \frac{1}{|B|}]$. Since $\varphi \simeq \psi$, $\varphi(y, \frac{t}{L}) \leqslant \psi(y, t) \leqslant \frac{1}{|B|}$. If $\varphi(y, \frac{t}{L}) \geqslant 1$, we obtain, by (A1$'$) of φ with argument $\frac{1}{L}t$,

$$\psi(x, \tfrac{\beta}{L^2}t) \leqslant \varphi(x, \tfrac{\beta}{L}t) \leqslant \varphi(y, \tfrac{1}{L}t) \leqslant \psi(y, t).$$

Otherwise, there exists $s \in [\frac{1}{L}, L]$ with $\varphi(y, st) = 1$, since $\varphi \in \Phi_s(\Omega)$ and $\varphi(y, Lt) \geqslant \psi(y, t) \geqslant 1$. Then

$$\psi(x, \tfrac{\beta}{L^2}t) \leqslant \varphi(x, \beta st) \leqslant \varphi(y, st) = 1 \leqslant \psi(y, t).$$

Thus ψ satisfies (A1$'$) with constant $\frac{\beta}{L^2}$.
(c) Suppose then that $\varphi \in \Phi_w(\Omega)$ satisfies (A0) and (A1$'$) and let t be such that $\psi(y, t) \in [1, \frac{1}{|B|}]$. As in (b), we obtain that $\psi(x, \frac{\beta}{L^2}t) \leqslant \psi(y, t)$ when $\varphi(y, \frac{t}{L}) \geqslant 1$. Assume then that $\varphi(y, \frac{t}{L}) < 1$. Since φ satisfies (A0), so does ψ and thus $\varphi(x, \beta_0^\varphi) \leqslant 1 \leqslant \varphi(x, 1/\beta_0^\varphi)$ and similarly for ψ. We obtain that

$\frac{t}{L} < \frac{1}{\beta_0^{\varphi}}$ and hence

$$\psi(x, \beta_0^{\varphi} \beta_0^{\psi} \tfrac{t}{L}) \leqslant \psi(x, \beta_0^{\psi}) \leqslant 1 \leqslant \psi(y, t).$$

Thus ψ satisfies (A1′) with constant $\min\{\frac{\beta}{L^2}, \beta_0^{\varphi} \beta_0^{\psi} \tfrac{1}{L}\}$.   □

Next we show that (A1) implies (A1′). For the other direction we need the additional assumption $\varphi \in \Phi_s(\Omega)$.

Question 4.1.4   Are (A1) and (A1′) equivalent for all $\varphi \in \Phi_w(\Omega)$?

Proposition 4.1.5

(a) *If $\varphi \in \Phi_w(\Omega)$ satisfies* (A1), *then it satisfies* (A1′).
(b) *If $\varphi \in \Phi_s(\Omega)$ satisfies* (A1′), *then it satisfies* (A1).

Proof

(a) Assume that (A1) holds and let $\psi \in \Phi_s(\Omega)$ with $\varphi \simeq \psi$ (Theorem 2.5.10). By Lemma 4.1.3(a), ψ satisfies (A1). Let $|B| \leqslant 1$, $x, y \in B \cap \Omega$ and $s \in [1, \frac{1}{|B|}]$. Then

$$\beta \psi^{-1}(x, s) \leqslant \psi^{-1}(y, s).$$

Since $\psi \in \Phi_s(\Omega)$, we can choose $t \in (0, \infty)$ such that $\psi(x, t) = s$. We apply $\psi(y, \cdot)$ to both sides of the inequality and use Lemma 2.3.3:

$$\psi(y, \beta \psi^{-1}(x, \psi(x, t))) \leqslant \psi(y, \psi^{-1}(y, \psi(x, t))) = \psi(x, t).$$

Since $\psi(x, t) = s \in [1, \frac{1}{|B|}]$, Corollary 2.3.4 implies that $\psi^{-1}(x, \psi(x, t)) = t$, and we have (A1′). Since $\psi \in \Phi_s(\Omega)$ and $\varphi \simeq \psi$, it follows from Lemma 4.1.3(b) that φ satisfies (A1′) as well.

(b) Assume that (A1′) holds with constant β. Let $|B| \leqslant 1$, $x, y \in B \cap \Omega$ and $s \in [1, \frac{1}{|B|}]$. Since $\varphi \in \Phi_s(\Omega)$, we can find t such that $\varphi(y, t) = s$. Then, by (A1′),

$$\varphi(x, \beta t) \leqslant \varphi(y, t).$$

Since $s \in [1, \frac{1}{|B|}]$, Corollary 2.3.4 implies that $\varphi^{-1}(y, s) = \varphi^{-1}(y, \varphi(y, t)) = t$. By Corollary 2.3.4, we obtain that

$$\beta \varphi^{-1}(y, s) = \beta t = \varphi^{-1}(x, \varphi(x, \beta t)) \leqslant \varphi^{-1}(x, \varphi(y, t)) = \varphi^{-1}(x, s)$$

provided that $\varphi(x, \beta t) \in (0, \infty)$. Since $\varphi(x, \beta t) \leqslant \varphi(y, t) = s \leqslant \frac{1}{|B|} < \infty$, we are left to prove the case $\varphi(x, \beta t) = 0$. Then $\beta t \leqslant \varphi^{-1}(x, t')$ for all $t' > 0$. Hence we obtain $\beta \varphi^{-1}(y, s) = \beta t \leqslant \varphi^{-1}(x, s)$. Thus (A1′) implies (A1).   □

Corollary 4.1.6 Let $\varphi \in \Phi_w(\Omega)$ satisfy (A0). Then φ satisfies (A1) if and only if it satisfies (A1').

Proof (A1) yields (A1') by Proposition 4.1.5(a). So let us assume that φ satisfies (A1'). Let $\psi \in \Phi_s(\Omega)$ with $\psi \simeq \varphi$ (Theorem 2.5.10). Since φ satisfies (A0), Lemma 4.1.3(c) yields that ψ satisfies (A1'), and hence by Proposition 4.1.5(b), ψ satisfies (A1). Thus by Lemma 4.1.3(a), φ satisfies (A1). □

Lemma 4.1.7 If $\varphi \in \Phi_w(\Omega)$ satisfies (A1), then φ^* satisfies (A1).

Proof By Theorem 2.4.8 and (A1) of φ we obtain

$$(\varphi^*)^{-1}(x, t) \approx \frac{t}{\varphi^{-1}(x, t)} \leqslant \frac{t}{\beta \varphi^{-1}(y, t)} \approx \frac{1}{\beta}(\varphi^*)^{-1}(y, t)$$

for $t \in [1, \frac{1}{|B|}]$, almost every $x, y \in B \cap \Omega$ and every ball B. □

When Ω is convex we can give a more flexible characterization of (A1).

Lemma 4.1.8 Let $\Omega \subset \mathbb{R}^n$ be convex, $\varphi \in \Phi_w(\Omega)$ and $0 < r \leqslant s$. Then φ satisfies (A1) if and only if there exists $\beta \in (0, 1)$ such that

$$\beta \varphi^{-1}(x, t) \leqslant \varphi^{-1}(y, t)$$

for every $t \in [r, \frac{s}{|B|}]$, almost every $x, y \in B \cap \Omega$ and every ball B.

Here, naturally, the constants β may be different in (A1) and in the condition.

Proof Let us first assume that φ satisfies (A1). Fix a ball B with $|B| \leqslant \frac{s}{r}$ and $x, y \in B$ and let $t \in [r, \frac{s}{|B|}]$. Let ℓ be a line segment connecting x to y. We cover ℓ by balls B_j with $|B_j| = \frac{r}{s}|B|$. Let R be the radius of B_j and $x_0 = x$. For $j = 1, 2, \ldots$ we choose points $x_j \in \ell$ such that $\ell(x_{j-1}, x_j) = \frac{R}{2}$ and finally set $x_k = y$. See Fig. 4.1. Note that k depends only on $\frac{r}{s}$.

Then $|B_j| \leqslant 1$ and $\frac{t}{r} \in [1, \frac{1}{|B_j|}]$. It follows from (A1) for each ball B_j that $\beta \varphi^{-1}(x_{j-1}, \frac{t}{r}) \leqslant \varphi^{-1}(x_j, \frac{t}{r})$ and hence

$$\beta^k \varphi^{-1}(x, \tfrac{t}{r}) \leqslant \varphi^{-1}(y, \tfrac{t}{r}).$$

If $r < 1$, then $\varphi^{-1}(x, \frac{t}{r}) \geqslant \varphi^{-1}(x, t)$ since φ^{-1} is increasing, and $(\mathrm{aDec})_1$ of φ^{-1} (Proposition 2.3.7) gives $\varphi^{-1}(y, \frac{t}{r}) \leqslant \frac{a}{r}\varphi^{-1}(y, t)$. Combining this with the previous inequality, we see that the condition holds. If $r \geqslant 1$, then increasing and $(\mathrm{aDec})_1$ are applied the other way around to reach the same conclusion.

Analogous steps also show that the condition implies (A1), so the proof is complete. □

Fig. 4.1 A chain of balls
from the proof of
Lemma 4.1.8

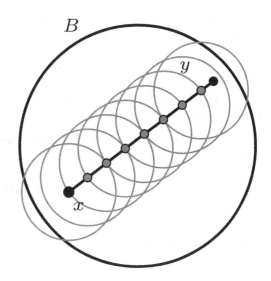

4.2 The Decay Condition (A2)

Definition 4.2.1 We say that $\varphi \in \Phi_w(\Omega)$ satisfies (A2) if for every $s > 0$ there exist $\beta \in (0, 1]$ and $h \in L^1(\Omega) \cap L^\infty(\Omega)$ such that

$$\beta \varphi^{-1}(x, t) \leqslant \varphi^{-1}(y, t)$$

for almost every $x, y \in \Omega$ and every $t \in [h(x) + h(y), s]$.

Since $\varphi \simeq \psi$ is equivalent to $\varphi^{-1} \approx \psi^{-1}$ (Theorem 2.3.6), we see that (A2) is invariant under equivalence.

Lemma 4.2.2 (A2) *is invariant under equivalence of weak Φ-functions.*

Furthermore, (A2) only concerns the behavior of φ at infinity, as the following result shows.

Lemma 4.2.3 *If $\Omega \subset \mathbb{R}^n$ is bounded, then every $\varphi \in \Phi_w(\Omega)$ satisfies* (A2).

Proof Choose $h := s\chi_\Omega$. Since Ω is bounded, $h \in L^1(\Omega)$. Further, the interval $[h(x) + h(y), s]$ is empty for every $x, y \in \Omega$, so (A2) holds vacuously.   □

Also the conjugate is nicely behaved.

Lemma 4.2.4 *If $\varphi \in \Phi_w(\Omega)$ satisfies* (A2), *then so does φ^*.*

Proof Suppose that

$$\beta \varphi^{-1}(x, t) \leqslant \varphi^{-1}(y, t)$$

for almost every $x, y \in \Omega$ and every $t \in [h(x)+h(y), s]$. We multiply the inequality by $(\varphi^*)^{-1}(x, t)(\varphi^*)^{-1}(y, t)$ and use Theorem 2.4.8:

$$ct(\varphi^*)^{-1}(y, t) \leqslant t(\varphi^*)^{-1}(x, t).$$

For $t > 0$ we divide the previous inequality by t to get (A2) of φ^*; for $t = 0$, (A2) is trivial, since $(\varphi^*)^{-1}(y, 0) = 0$. $\qquad \square$

Lemma 4.2.5 *The Φ-function $\varphi \in \Phi_w(\Omega)$ satisfies (A2) if and only if for every $s > 0$ there exist $\beta \in (0, 1]$ and $h \in L^1(\Omega) \cap L^\infty(\Omega)$ such that*

$$\varphi(x, \beta t) \leqslant \varphi(y, t) + h(x) + h(y)$$

for almost every $x, y \in \Omega$ whenever $\varphi(y, t) \in [0, s]$.

Proof We first show that the condition in the lemma is invariant under equivalence of Φ-functions. Let $\varphi \simeq \psi \in \Phi_w(\Omega)$ with constant $L \geqslant 1$. Then

$$\psi(x, \tfrac{\beta}{L}t) \leqslant \varphi(x, \beta t) \leqslant \varphi(y, t) + h(x) + h(y) \leqslant \psi(y, Lt) + h(x) + h(y),$$

for $\varphi(y, t) \in [0, s]$. Denote $t' := Lt$. If $\psi(y, t') \in [0, s]$, then $\varphi(y, t) \in [0, s]$ and the previous inequality gives $\psi(x, \tfrac{\beta}{L^2}t') \leqslant \psi(y, t') + h(x) + h(y)$, which is the condition for ψ. By Lemma 4.2.2, (A2) is invariant under equivalence of weak Φ-functions. Hence it suffices to show the claim for $\varphi \in \Phi_s(\Omega)$, by Theorem 2.5.10.

Assume (A2) and denote $\tau := \varphi^{-1}(x, t)$. By Lemma 2.3.3, $\varphi(x, \tau) = t$, and it follows from (A2) that

$$\beta_2\tau \leqslant \varphi^{-1}(y, \varphi(x, \tau))$$

for almost every $x, y \in \Omega$ whenever $\varphi(x, \tau) \in [h(x) + h(y), s]$. Then we apply $\varphi(y, \cdot)$ to both sides and use Lemma 2.3.3 to obtain that

$$\varphi(y, \beta_2\tau) \leqslant \varphi(x, \tau)$$

for the same range. If, on the other hand, $\varphi(x, \tau) \in [0, h(x) + h(y))$, then we can find $\tau' > \tau$ such that $\varphi(x, \tau') = h(x) + h(y)$ since $\varphi \in \Phi_s(\Omega)$. As $\varphi(x, \cdot)$ is increasing, we obtain by the previous case applied for τ' that

$$\varphi(y, \beta_2\tau) \leqslant \varphi(y, \beta_2\tau') \leqslant \varphi(x, \tau') = h(x) + h(y).$$

Combining the two cases, we conclude that

$$\varphi(y, \beta_2\tau) \leqslant \varphi(x, \tau) + h(x) + h(y)$$

whenever $\varphi(x, \tau) \in [0, s]$, which is the condition.

Assume conversely that

$$\varphi(y, \beta_2 \tau) \leqslant \varphi(x, \tau) + h(x) + h(y),$$

for $\varphi(x, \tau) \in [0, s]$. By adding $(1 + |x|)^{-(n+1)}$ to h, we may assume that $h > 0$ everywhere. By $(\text{aInc})_1$, we conclude that

$$\varphi(y, \tfrac{\beta_2}{2a}\tau) \leqslant \tfrac{1}{2}\varphi(y, \beta_2 \tau) \leqslant \tfrac{\varphi(x,\tau)+h(x)+h(y)}{2} \leqslant \max\{\varphi(x, \tau), h(x) + h(y)\}.$$

If $\varphi(x, \tau) \in [h(x) + h(y), s]$, then this implies that

$$\varphi(y, \tfrac{\beta_2}{2a}\tau) \leqslant \varphi(x, \tau) =: t.$$

Note by Corollary 2.3.4 that $\tau = \varphi^{-1}(x, t)$. Next we apply $\varphi^{-1}(y, \cdot)$ to both sides of the previous inequality and use Corollary 2.3.4:

$$\tfrac{\beta_2}{2a}\varphi^{-1}(x, t) = \varphi^{-1}(y, \varphi(y, \tfrac{\beta_2}{2a}\varphi^{-1}(x, t))) = \varphi^{-1}(y, \varphi(y, \tfrac{\beta_2}{2a}\tau)) \leqslant \varphi^{-1}(y, t)$$

provided that $\varphi(y, \tfrac{\beta_2}{2a}\varphi^{-1}(x, t)) \in (0, \infty)$. Since $\varphi(y, \tfrac{\beta_2}{2a}\varphi^{-1}(x, t)) \leqslant \varphi(x, \tau) \leqslant s < \infty$, we are left to show the claim in the case $\varphi(y, \tfrac{\beta_2}{2a}\varphi^{-1}(x, t)) = 0$. But then it follows from the definition of the inverse that $\tfrac{\beta_2}{2a}\varphi^{-1}(x, t) \leqslant \varphi^{-1}(y, t')$ for all $t' > 0$, so this holds in particular for $t' = t > 0$. Thus we have shown (A2). $\qquad\square$

A problem with (A2) and the condition in Lemma 4.2.5 is that we have h at two points, $h(x)$ and $h(y)$. Next we consider a condition where there is only one "moving point". We will later use it to verify (A2) in special cases in Chap. 7.

Definition 4.2.6 We say that $\varphi \in \Phi_w(\Omega)$ satisfies (A2′) if there exist $\varphi_\infty \in \Phi_w$, $h \in L^1(\Omega) \cap L^\infty(\Omega)$, $s > 0$ and $\beta \in (0, 1]$ such that

$$\varphi(x, \beta t) \leqslant \varphi_\infty(t) + h(x) \quad \text{and} \quad \varphi_\infty(\beta t) \leqslant \varphi(x, t) + h(x)$$

for almost every $x \in \Omega$ when $\varphi_\infty(t) \in [0, s]$ and $\varphi(x, t) \in [0, s]$, respectively.

Suppose that h in the definition satisfies $\lim_{|x| \to \infty} |h(x)| = 0$. Then

$$\varphi_\infty^+(t) := \limsup_{|x| \to \infty} \varphi(x, t) \simeq \liminf_{|x| \to \infty} \varphi(x, t) =: \varphi_\infty^-(t).$$

By Lemma 2.5.18, $\varphi_\infty^\pm \in \Phi_w(\Omega)$ provided that they are non-degenerate; this holds for example if (A0) holds. In this case we may take either of these functions for φ_∞. In this sense φ_∞ is the limit of φ as $x \to \infty$.

Lemma 4.2.7 (A2) *and* (A2′) *are equivalent.*

Proof We start by showing that (A2′) implies (A2). Let $s' > 0$ be from (A2′). Suppose that $\varphi(y, t) \in [0, s']$. Then $\varphi_\infty(\beta t) \leqslant \varphi(y, t) + h(y)$. Denote $\beta' := (2a(1+$

$\frac{1}{s'}\|h\|_\infty))^{-1}$. Since φ_∞ satisfies (aInc)$_1$,

$$\varphi_\infty(\beta\beta' t) \leqslant \frac{1}{2(1+\frac{1}{s'}\|h\|_\infty)}\varphi_\infty(\beta t) \leqslant \tfrac{1}{2}\varphi(y,t) + \frac{s'}{2\|h\|_\infty}h(y) \leqslant s'.$$

Hence by the other inequality of the assumption,

$$\varphi(x, \beta^2\beta' t) \leqslant \varphi_\infty(\beta\beta' t) + h(x) \leqslant \varphi(y,t) + h(y) + h(x)$$

for $\varphi(y, t) \in [0, s']$. Let $s > s'$ and $\varphi(y,t) \in [0, s]$. Then, by (aInc)$_1$, $\varphi(y, \frac{s't}{as}) \leqslant a\frac{s'}{as}\varphi(y, t) \leqslant s'$ and hence by the previous case

$$\varphi(x, \beta^2\beta'\tfrac{s'}{as}t) \leqslant \varphi(y, \tfrac{s'}{as}t) + h(y) + h(x) \leqslant \varphi(y, t) + h(y) + h(x).$$

Now (A2) follows by Lemma 4.2.5.

Next we show that (A2) implies (A2′). Let h, s and β be as in Lemma 4.2.5 and let N be a set of zero measure such that $h|_{\Omega\setminus N} \leqslant \|h\|_\infty$ and

$$\varphi(x, \beta t) \leqslant \varphi(y, t) + h(x) + h(y) \tag{4.2.1}$$

for every $x, y \in \Omega \setminus N$ whenever $\varphi(y, t) \in [0, s]$. In the following we work only with points from $\Omega \setminus N$, which is sufficient since we need to establish the claim almost everywhere.

Choose a sequence $x_i \to \infty$ for which $h(x_i) \to 0$. We define

$$\varphi_\infty(t) := \limsup_{i\to\infty} \varphi(x_i, t).$$

Using this in (4.2.1) for x, we obtain

$$\varphi_\infty(\beta t) \leqslant \varphi(y, t) + h(y)$$

for every $y \in \Omega \setminus N$ and every $\varphi(y, t) \in [0, s]$. Assume next that $\varphi_\infty(t) \in [0, s/2]$. Then $\varphi(x_i, t) \in [0, s]$ for large i. Hence by (4.2.1) we have

$$\varphi(x, \beta t) \leqslant \varphi(x_i, t) + h(x) + h(x_i) \to \varphi_\infty(t) + h(x)$$

for $x \in \Omega \setminus N$. Thus (A2′) holds in $[0, \frac{s}{2}]$ once we show that $\varphi_\infty \in \Phi_w$.

Each $\varphi(x_i, \cdot)$ is increasing and satisfies (aInc)$_1$ with uniform constant, so these properties are inherited by φ_∞. Fix $t > 0$ with $\varphi(y, t) \leqslant s$. By (4.2.1),

$$\varphi(x_i, \beta t) \leqslant \varphi(y, t) + h(x_i) + h(y) \leqslant \varphi(y, t) + 2\|h\|_\infty.$$

Hence $\varphi_\infty(\beta t) < \infty$ and so it follows from (aInc)$_1$ that $\lim_{t\to 0}\varphi_\infty(t) = 0$.

If $\varphi_\infty(t) > 0$ for some $t > 0$, then we similarly deduce that $\lim_{t\to\infty} \varphi_\infty(t) = \infty$. Otherwise, we would have $\varphi_\infty(t) \equiv 0$, which we show is impossible. Choose T such that $\varphi(y, T) > \|h\|_\infty$ for some fixed $y \in \Omega$. Then

$$\varphi(y, T) \leqslant \varphi(x_i, \tfrac{1}{\beta}T) + h(x_i) + h(y)$$

for all sufficiently large i, since $\varphi(x_i, \tfrac{1}{\beta}T) \in [0, s]$. Taking lim sup implies that $\varphi(y, T) \leqslant \|h\|_\infty$, a contradiction. Hence this case cannot occur, so we have established that φ_∞ is a weak Φ-function.   □

Question 4.2.8 Is (A0) needed for the equivalence in the next result?

Lemma 4.2.9 *Let $\varphi \in \Phi_w(\Omega)$ and let us consider the following conditions:*

(a) *The Φ-function φ satisfies* (A2$'$).
(b) *There exist $\varphi_\infty \in \Phi_w$, $h \in L^1(\Omega) \cap L^\infty(\Omega)$ and $\beta \in (0, 1]$ such that*

$$\varphi(x, \beta t) \leqslant \varphi_\infty(t) + h(x) \quad and \quad \varphi_\infty(\beta t) \leqslant \varphi(x, t) + h(x)$$

when $\varphi_\infty(t) \in [0, 1]$ and $\varphi(x, t) \in [0, 1]$ respectively.
(c) *$L^\varphi(\Omega) \cap L^\infty(\Omega) = L^{\varphi_\infty}(\Omega) \cap L^\infty(\Omega)$ with equivalent norms.*

Then (a) *and* (b) *are equivalent. And if, additionally, φ and φ_∞ satisfy* (A0), *then* (a), (b) *and* (c) *are equivalent.*

Proof Let us first show that (a) and (b) are equivalent. Clearly (b) implies (a). So let us assume that (A2$'$) holds for some $s < 1$. Let $\varphi(y, t) \in [0, 1]$. Then by (aInc)$_1$ $\varphi(y, \tfrac{st}{a}) \leqslant a\tfrac{s}{a}\varphi(y, t) \leqslant s$ and hence by (A2$'$)

$$\varphi(x, \beta\tfrac{s}{a}t) \leqslant \varphi(y, \tfrac{s}{a}t) + h(y) + h(x) \leqslant \varphi(y, t) + h(y) + h(x).$$

And similarly for the other inequality.

Let us then show that (b) and (c) are equivalent. By (A0) there exists $\beta_0 > 0$ such that $\varphi(x, \beta_0) \leqslant 1 \leqslant \varphi(x, \tfrac{1}{\beta_0})$ and $\varphi_\infty(\beta_0) \leqslant 1 \leqslant \varphi_\infty(\tfrac{1}{\beta_0})$. We define $\xi \in \Phi_w(\Omega)$ and $\psi \in \Phi_w$ by

$$\xi(x, t) := \max\{\varphi(x, t), \infty\chi_{(\beta_0,\infty)}(t)\} \quad and \quad \psi(t) := \max\{\varphi_\infty(t), \infty\chi_{(\beta_0,\infty)}(t)\}.$$

Then $L^\varphi \cap L^\infty = L^\xi$ and $L^{\varphi_\infty} \cap L^\infty = L^\psi$ so that (c) becomes $L^\xi = L^\psi$. Suppose that (b) holds. If $t \leqslant \beta_0$, then $\varphi(x, t) \leqslant 1$ and (b) implies that $\varphi_\infty(\beta t) \leqslant \varphi(x, t) + h(x)$. If $t > \beta_0$, then $\psi(t) = \infty$. Hence

$$\xi(x, \beta t) \leqslant \psi(t) + h(x)$$

for all $t \geqslant 0$. Similarly we prove that $\psi(\beta t) \leqslant \xi(x, t) + h(x)$. Then $L^\xi = L^\psi$ by Theorem 3.2.6. It is well-known that $L^\xi = L^\psi$ if and only if the norms are equivalent. For completeness, this is shown in Remark 4.2.10.

Assume now (c) and define

$$\xi(x,t) := \max\{\varphi(x,t), \infty\chi_{(\frac{1}{\beta_0},\infty)}(t)\} \quad \text{and} \quad \psi(t) := \max\{\varphi_\infty(t), \infty\chi_{(\frac{1}{\beta_0},\infty)}(t)\}.$$

Still (c) is equivalent to $L^\xi = L^\psi$ and Theorem 3.2.6 implies that there exist $\beta > 0$ and $h \in L^1$ such that

$$\xi(x,\beta t) \leqslant \psi(t) + h(x) \quad \text{and} \quad \psi(\beta t) \leqslant \xi(x,t) + h(x).$$

When $\varphi_\infty(t) \in [0,1]$, we have $t \leqslant \frac{1}{\beta_0}$ and so the first inequality gives

$$\xi(x,\beta t) \leqslant \varphi_\infty(t) + h(x).$$

Since the right-hand side is finite almost everywhere, $\xi(x,\beta t) = \varphi(x,\beta t)$ and so we get $\varphi(x,\beta t) \leqslant \varphi_\infty(t) + h(x)$. Furthermore, we may replace h by $\min\{h,1\}$ since $\varphi(x,\beta t) \leqslant 1$ provided $\beta \leqslant \beta_0^2$. This gives one of the inequalities in (b); the other is proved analogously, starting from the inequality with $\psi(\beta t)$.   □

Remark 4.2.10 In (3.3.2) we noted that L^φ is solid. In solid Banach spaces A and B, the equality of sets $A = B$ implies also the equivalence of norms as the following argument shows. If $\|f\|_A \not\lesssim \|f\|_B$, then we can choose f_i such that $\|f_i\|_A \geqslant 3^i$ but $\|f_i\|_B \leqslant 1$. Now for $g := \sum_i 2^{-i}|f_i|$ we have

$$\|g\|_A \geqslant \|2^{-i}f_i\|_A \geqslant (3/2)^i \to \infty$$

and $\|g\|_B \leqslant \sum_i 2^{-i} = 1$ so that $g \in B \setminus A$. Hence $A \neq B$. The implication $A = B \Rightarrow \|\cdot\|_A \approx \|\cdot\|_B$ follows by contraposition. The same argument works also in quasi-Banach spaces provided that their quasinorms satisfy countable subadditivity $\|\sum_i f_i\| \lesssim \sum_i \|f_i\|$, see Corollary 3.2.5.

Consider the example $\varphi(x,t) := \frac{1}{1+|x|}(t-1)_+$. We use it to show that the next result is not true without the assumption (A0). Let φ_∞ be from (A2′). Choose $t \geqslant 2$ such that $c := \varphi_\infty(\beta t) > 0$. When $|x|$ is large enough, we have $\varphi(x,t) \in [0,s]$ and hence

$$c = \varphi_\infty(\beta t) \leqslant \varphi(x,t) + h(x) = \frac{1}{1+|x|}(t-1)_+ + h(x).$$

This yields that $\liminf_{|x|\to\infty} h(x) > c$ and hence $h \notin L^1$ so (A2′) can not hold.

Corollary 4.2.11 *Let* $\varphi \in \Phi_w(\Omega)$ *satisfy* (A0). *If there exists* $\tau > 0$ *such that* $\varphi(x,t)$ *is independent of* x *for* $t \in [0,\tau]$, *then* φ *satisfies* (A2).

Proof Assume that $\varphi(x,t)$ is independent of x for $t \in [0,\tau]$ and that $\varphi(x,\beta_0) \leqslant 1 \leqslant \varphi(x,\frac{1}{\beta_0})$ for almost all $x \in \Omega$. Define $\varphi_\infty(t) := \varphi(x_0,t) + (t-\tau)_+$ for some fixed $x_0 \in \Omega$.

We choose $s := \frac{1}{2}$, $\beta := \min\{\tau\beta_0, \frac{\tau}{s+\tau}\}$ and $h := 0$ in (A2′). If $\varphi(x, t) \in [0, s]$, then $t \leqslant \frac{1}{\beta_0}$. Hence $\beta t \in [0, \tau]$ so that

$$\varphi(x, \beta t) = \varphi_\infty(\beta t) \leqslant \varphi_\infty(t).$$

On the other hand, if $\varphi_\infty(t) \in [0, s]$, then $t \leqslant s + \tau$. Hence $\beta t \in [0, \tau]$ so that

$$\varphi_\infty(\beta t) = \varphi(x, \beta t) \leqslant \varphi(x, t).$$

Thus (A2′) holds so the claim follows from Lemma 4.2.7. $\qquad\qquad\square$

4.3 Maximal Operators

For $f \in L^1_{\mathrm{loc}}(\Omega)$ we define

$$Mf(x) := \sup_{B \ni x} \frac{1}{|B|} \int_{B \cap \Omega} |f(y)| \, dy$$

and call M the (non-centered Hardy–Littlewood) *maximal operator*, where the supremum is taken over all open balls containing x. Often, one uses the centered maximal operator $M_\odot$, where the supremum is taken over all open balls centered at x, which is defined by

$$M_\odot f(x) := \sup_{r > 0} \frac{1}{|B(x, r)|} \int_{B(x,r) \cap \Omega} |f(y)| \, dy.$$

Note that the centered and non-centered maximal operators are point-wise equivalent: for every $f \in L^1_{\mathrm{loc}}(\mathbb{R}^n)$

$$2^{-n} Mf \leqslant M_\odot f \leqslant Mf. \qquad (4.3.1)$$

Recall, that for $\varphi \in \Phi_{\mathrm{w}}(\Omega)$, we defined

$$\varphi_B^-(t) := \operatorname*{ess\,inf}_{x \in B \cap \Omega} \varphi(x, t) \quad \text{and} \quad \varphi_B^+(t) := \operatorname*{ess\,sup}_{x \in B \cap \Omega} \varphi(x, t)$$

in Sect. 2.5. Note that in the next lemma we consider φ independent of x, e.g. it could be $\varphi_B^\pm$ provided it is non-degenerate.

Lemma 4.3.1 *Let $\varphi : [0, \infty) \to [0, \infty]$ be a Φ-prefunction that satisfies* (aInc)$_p$, *$p > 0$. Then there exists $\beta > 0$ such that the following Jensen-type inequality holds* *for every measurable set* U, $|U| \in (0, \infty)$, *and every* $f \in L^1(U)$:

$$\varphi\left(\beta \fint_U |f|\, dx\right)^{\frac{1}{p}} \leqslant \fint_U \varphi(f)^{\frac{1}{p}}\, dx.$$

Proof Since $\varphi^{\frac{1}{p}}$ satisfies (aInc)$_1$, there exists $\psi \simeq \varphi^{\frac{1}{p}}$ which is convex, by Lemma 2.2.1. By Jensen's inequality for ψ,

$$\varphi\left(\frac{1}{L^2}\fint_U |f|\, dx\right)^{\frac{1}{p}} \leqslant \psi\left(\fint_U \tfrac{1}{L}|f|\, dx\right) \leqslant \fint_U \psi(\tfrac{1}{L}f)\, dx \leqslant \fint_U \varphi(f)^{\frac{1}{p}}\, dx.$$

□

The following is the key estimate for boundedness and weak type estimates of maximal and averaging operators. It is the counterpart of the key-estimate Theorem 4.2.4 in our previous monograph [34]. The later formulation is needed for the modular forms of the Poincaré (Proposition 6.2.10) and Sobolev–Poincaré (Corollary 6.3.13) inequalities.

Theorem 4.3.2 (Key Estimate) *Let* $\varphi \in \Phi_w(\Omega)$ *satisfy* (A0), (A1), (A2) *and* (aInc)$_p$, $p \in [1, \infty)$. *Then there exist* $\beta > 0$ *and* $h \in L^1(\Omega) \cap L^\infty(\Omega)$ *such that*

$$\varphi\left(x, \frac{\beta}{|B|}\int_{B\cap\Omega} |f|\, dy\right)^{\frac{1}{p}} \leqslant \frac{1}{|B|}\int_{B\cap\Omega} \varphi(y, |f|)^{\frac{1}{p}}\, dy + h(x)^{\frac{1}{p}} + \frac{1}{|B|}\int_{B\cap\Omega} h(y)^{\frac{1}{p}}\, dy$$

and

$$\varphi\left(x, \frac{\beta}{|B|}\int_{B\cap\Omega} |f|\, dy\right)^{\frac{1}{p}} \leqslant \frac{1}{|B|}\int_{B\cap\Omega} \varphi(y, |f|)^{\frac{1}{p}}\, dy + \|h\|_\infty^{\frac{1}{p}} \frac{|\{f \neq 0\} \cap B \cap \Omega|}{|B|}$$

for every ball B, $x \in B \cap \Omega$ *and* $f \in L^\varphi(\Omega)$ *with* $\varrho_\varphi(f) \leqslant 1$.

Proof Let us start from the first claim. Let β_0 be the constant from (A0), β_1 the constant from (A1$'$) that follows from (A1) by Proposition 4.1.5, and β_2 from Lemma 4.2.5 with $s = 1$ where we used (A2). Let β_J be the constant from the Jensen-type inequality Lemma 4.3.1.

We may assume without loss of generality that $f \geqslant 0$. Fix a ball B, denote $\hat{B} := B \cap \Omega$ and choose $x \in \hat{B}$. Denote $f_1 := f\chi_{\{f > 1/\beta_0\}}$, $f_2 := f - f_1$, and $A_i := \frac{1}{|B|}\int_{\hat{B}} f_i\, dy$. Since $\varphi^{1/p}$ is increasing,

$$\varphi\left(x, \frac{\beta}{|B|}\int_{\hat{B}} f\, dy\right)^{\frac{1}{p}} \leqslant \varphi\left(x, 2\beta \max\{A_1, A_2\}\right)^{\frac{1}{p}} \leqslant \varphi\left(x, 2\beta A_1\right)^{\frac{1}{p}} + \varphi\left(x, 2\beta A_2\right)^{\frac{1}{p}}$$

where β is a constant that will be fixed later.

We first estimate A_1. By Lemmas 2.5.16 and 4.3.1 (with $p = 1$) and $\varrho_\varphi(f\chi_{\{|f|>1/\beta_0\}}) \leqslant \varrho_\varphi(f) \leqslant 1$ we obtain that

$$\varphi_B^-(\beta_J A_1) \leqslant \frac{1}{|B|} \int_{\hat B} \varphi_B^-(f_1)\,dy \leqslant \frac{1}{|B|} \int_{\hat B} \varphi(y, f_1)\,dy \leqslant \frac{1}{|B|}.$$

Note that in fact instead of $\varrho_\varphi(f) \leqslant 1$ we need to assume only that $\varrho_\varphi(f_1) \leqslant 1$. It follows by (aInc)$_1$ that $\varphi_B^-\left(\frac{1}{2a}\beta_J A_1\right) < \frac{1}{|B|}$. Denote $\beta_J' := \frac{1}{2a}\beta_J$.

Suppose first that $\varphi_B^-(\beta_J' A_1) \geqslant 1$. Then there exists $y \in \hat B$ with $\varphi(y, \beta_J' A_1) \in [1, \frac{1}{|B|}]$ and $\varphi(y, \beta_J' A_1) \leqslant 2\varphi_B^-(\beta_J' A_1)$. By Proposition 4.1.5, φ satisfies (A1'), and so by Lemma 4.3.1 for the second inequality we obtain

$$\varphi(x, \beta_1\beta_J' A_1)^{\frac{1}{p}} \leqslant 2\varphi_B^-(\beta_J' A_1)^{\frac{1}{p}} \leqslant \frac{2}{|B|} \int_{\hat B} \varphi_B^-(f_1)^{\frac{1}{p}}\,dy \leqslant \frac{2}{|B|} \int_{\hat B} \varphi(y, f_1)^{\frac{1}{p}}\,dy.$$

Next consider the case $\varphi_B^-(\beta_J' A_1) < 1$. By (A0), this implies that $\beta_J' A_1 \leqslant \frac{1}{\beta_0}$. By (aInc)$_p$ and (A0), we conclude that

$$\varphi(x, \beta_0^2\beta_J' A_1)^{\frac{1}{p}} \leqslant a^{\frac{1}{p}} \beta_0\beta_J' A_1 \varphi(x, \beta_0)^{\frac{1}{p}} \leqslant a^{1/p}\beta_0\beta_J' A_1.$$

By (A0), $1 \leqslant \varphi(y, 1/\beta_0)$. Since $f_1 > 1/\beta_0$ when it is non-zero, we find, by (aInc)$_p$ for the second inequality, that

$$A_1 = \frac{1}{|B|} \int_{\hat B} f_1\,dy \leqslant \frac{1}{|B|} \int_{\hat B} f_1\varphi(y, \tfrac{1}{\beta_0})^{\frac{1}{p}}\,dy \leqslant \frac{a^{1/p}}{\beta_0|B|} \int_{\hat B} \varphi(y, f_1)^{\frac{1}{p}}\,dy.$$

In view of this and the conclusion of previous paragraph, we obtain the final estimate for f_1

$$\varphi\left(x, \frac{\beta_0}{2a^{2/p}}\beta_0^2\beta_1\beta_J' A_1\right)^{\frac{1}{p}} \leqslant \frac{\beta_0}{2a^{1/p}}\varphi(x, \beta_0^2\beta_1\beta_J' A_1)^{\frac{1}{p}} \leqslant \frac{1}{|B|} \int_{\hat B} \varphi(y, f_1)^{\frac{1}{p}}\,dy,$$

where we also used (aInc)$_p$ for the first inequality.

Let us move to f_2. Since $f_2 \leqslant \frac{1}{\beta_0}$ we obtain by (A0) that $\varphi(y, \beta_0^2 f_2) \leqslant \varphi(y, \beta_0) \leqslant 1$. We use Lemma 4.3.1 for $\varphi(x, \cdot)^{\frac{1}{p}}$ (with x fixed) and Lemma 4.2.5:

$$\varphi\left(x, \beta_J\beta_0^2\beta_2 A_2\right)^{\frac{1}{p}} \leqslant \frac{1}{|B|} \int_{\hat B} \varphi(x, \beta_0^2\beta_2 f_2(y))^{\frac{1}{p}}\,dy$$

$$\leqslant \frac{1}{|B|} \int_{\hat B} \varphi(y, \beta_0^2 f_2)^{\frac{1}{p}} + h(x)^{\frac{1}{p}} + h(y)^{\frac{1}{p}}\,dy \qquad (4.3.2)$$

$$\leqslant \frac{1}{|B|} \int_{\hat B} \varphi(y, f_2)^{\frac{1}{p}} + h(x)^{\frac{1}{p}} + h(y)^{\frac{1}{p}}\,dy.$$

Adding the estimates for f_1 and f_2, we conclude the proof of the first claim by choosing $\beta := \frac{1}{2}\min\{2^{-1}a^{-2/p}\beta_0^3\beta_1\beta_J', \beta_J\beta_0^2\beta_2\}$.

Let us then prove the second claim. For f_1 the proof is the same as in the previous case. Let us move to f_2. We proceed as in (4.3.2), but for the second inequality we use instead

$$\varphi(x, \beta_0^2\beta_2 f_2(y))^{\frac{1}{p}} \leqslant \chi_{\{f_2>0\}}(y)\big(\varphi(y, \beta_0^2 f_2)^{\frac{1}{p}} + 2\|h\|_\infty^{\frac{1}{p}}\big),$$

and conclude in the same way as before.   □

Taking the supremum over balls B in the previous lemma, and noticing that $h(x)^{1/p} \leqslant M(h^{1/p})(x)$, we obtain the following corollary:

Corollary 4.3.3 *Let $\varphi \in \Phi_w(\Omega)$ satisfy (A0), (A1), (A2) and (aInc)$_p$. Then there exists $\beta > 0$ and $h \in L^1(\Omega) \cap L^\infty(\Omega)$ such that*

$$\varphi(x, \beta Mf(x))^{\frac{1}{p}} \lesssim M\big(\varphi(\cdot, f)^{\frac{1}{p}}\big)(x) + M(h^{\frac{1}{p}})(x)$$

for every ball B, $x \in B \cap \Omega$ and $f \in L^\varphi(\Omega)$ with $\varrho_\varphi(f) \leqslant 1$.

The operator T is *bounded* from $L^\varphi(\Omega)$ to $L^\psi(\Omega)$ if $\|Tf\|_{L^\psi(\Omega)} \lesssim \|f\|_{L^\varphi(\Omega)}$ for all $f \in L^\varphi(\Omega)$. The next theorem shows that maximal operator is bounded provided that φ satisfies (A0)–(A2) and (aInc). Note that (aInc) ensures that the space is not of L^1-type.

Theorem 4.3.4 (Maximal Operator Estimate) *Let $\varphi \in \Phi_w(\Omega)$ satisfy (A0), (A1), (A2) and (aInc). Then the maximal operator is bounded on $L^\varphi(\Omega)$:*

$$M : L^\varphi(\Omega) \to L^\varphi(\Omega).$$

Proof Let φ satisfy (aInc)$_p$ with $p > 1$. Let $f \in L^\varphi(\Omega)$ be non-negative and choose $\varepsilon := \frac{1}{2\|f\|_\varphi}$; we may assume that $\|f\|_\varphi > 0$ since otherwise $f = 0$ and there is nothing to prove. Then $\varrho_\varphi(\varepsilon f) \leqslant 1$ by the unit ball property (Lemma 3.2.3). By Corollary 4.3.3 for the function εf,

$$\varphi(x, \beta\varepsilon Mf(x))^{\frac{1}{p}} \lesssim M\big(\varphi(\cdot, \varepsilon f)^{\frac{1}{p}}\big)(x) + M(h^{\frac{1}{p}})(x).$$

Raising both side to the power p and integrating, we find that

$$\int_\Omega \varphi(x, \beta\varepsilon Mf(x))\, dx \lesssim \int_\Omega \big[M\big(\varphi(\cdot, \varepsilon f)^{\frac{1}{p}}\big)(x)\big]^p\, dx + \int_\Omega \big[M\big(h^{\frac{1}{p}}\big)(x)\big]^p\, dx.$$

Since M is bounded on $L^p(\Omega)$, we obtain that

$$\int_\Omega \varphi(x, \beta\varepsilon Mf(x))\, dx \lesssim \int_\Omega \varphi(x, \varepsilon f)^{\frac{1}{p}p}\, dx + \int_\Omega h(x)^{\frac{1}{p}p}\, dx = \varrho_\varphi(\varepsilon f) + \|h\|_1.$$

Hence $\varrho_\varphi(\beta\varepsilon Mf) \leqslant c(1 + \|h\|_1) =: c_1$ and by (aInc)$_1$, $\varrho_\varphi(\frac{\beta\varepsilon}{ac_1} Mf) \leqslant 1$. By the unit ball property we have $\|\frac{\beta\varepsilon}{ac_1} Mf\|_\varphi \leqslant 1$ so that

$$\|Mf\|_\varphi \leqslant \frac{ac_1}{\beta\varepsilon} = \frac{2ac_1}{\beta} \|f\|_\varphi. \qquad \square$$

In some sense, (A1) is also a necessary assumption for boundedness of the maximal operator, as the next result shows.

Proposition 4.3.5 *Let $\varphi \in \Phi_w(\mathbb{R})$ be such that $x \mapsto \varphi(x, t)$ is increasing and $M : L^\varphi(\mathbb{R}) \to L^\varphi(\mathbb{R})$ is bounded. Then φ satisfies* (A1).

Proof Let $-\infty < x < y < \infty$ and $r \geqslant |x - y|$. Then

$$\varrho_\varphi(\tfrac{1}{\lambda}\chi_{[x-r,x]}) = \int_{x-r}^x \varphi(z, \tfrac{1}{\lambda}) \, dz \leqslant r\varphi(x, \tfrac{1}{\lambda}).$$

Therefore, by Lemma 2.3.9(a) and the definition of the norm,

$$\|\chi_{[x-r,x]}\|_\varphi \leqslant \frac{2}{\varphi^{-1}(x, \tfrac{1}{r})}.$$

On the other hand, $M(\chi_{[x-r,x]}) \geqslant \tfrac{1}{3}\chi_{[y,y+r]}$. Thus a calculation similar to the one above yields that

$$\|M(\chi_{[x-r,x]})\|_\varphi \geqslant \frac{1}{\varphi^{-1}(y, \tfrac{3}{r})}.$$

Then it follows from the boundedness of the maximal operator and (aDec)$_1$ of φ^{-1} that

$$\varphi^{-1}(x, \tfrac{1}{r}) \lesssim \varphi^{-1}(y, \tfrac{3}{r}) \lesssim \varphi^{-1}(y, \tfrac{1}{r}).$$

Since we may choose $\frac{1}{r} \in (0, \frac{1}{|x-y|}]$, this implies (A1). $\qquad \square$

Note that the constant exponent weighted case shows that (A0) is not necessary and examples in the variable exponent case [77] show that (A2$'$) is not necessary. In [34, Theorem 4.7.1], we showed that (aInc) is necessary in the variable exponent case. This is still open for the more general case.

Question 4.3.6 If the maximal operator is bounded on $L^\varphi(\mathbb{R}^n)$ does it imply that (aInc) holds?

Let us mention here an open problem related to the work of Diening [33] from 2005. He proved in the variable exponent case that the maximal operator is bounded on $L^{p(\cdot)}(\mathbb{R}^n)$ if and only if it is bounded on $L^{p'(\cdot)}(\mathbb{R}^n)$ (provided

$1 < p^- \leqslant p^+ < \infty$). Does this result generalize to the generalized Orlicz case? (Note that Diening already produced some partial results in [33].)

Question 4.3.7 Let $\varphi \in \Phi_w(\Omega)$ satisfy (aInc) and (aDec). Is it true that the maximal operator is bounded on $L^\varphi(\mathbb{R}^n)$ if and only if it is bounded on $L^{\varphi^*}(\mathbb{R}^n)$?

The maximal operator is a very strong tool in analysis. However, using the maximal estimate does incur a slight penalty: one has to assume (aInc). This is so already in the classical case, where the maximal operator is bounded on L^p if and only if the exponent $p \in (1, \infty]$. For instance if we want to prove the boundedness of convolution, this is easily done with the maximal operator, but not for L^1.

Next we consider two alternatives which enable us to obtain further results without the extraneous assumption (aInc): weak type estimates of the maximal operator and strong type estimates of the averaging operator.

We need the following covering theorem. See [40, Theorem 2, p. 30] or [91, Theorem 2.7, p. 30] for the proof.

Theorem 4.3.8 (Besicovitch Covering Theorem) *Let A be a bounded set in $\mathbb{R}^n$. For each $x \in A$ a cube (or ball) $Q_x \subset \mathbb{R}^n$ centered at x is given. Then one can choose, from among the given sets $\{Q_x\}_{x \in A}$ a sequence $\{Q_j\}_{j \in \mathbb{N}}$ (possibly finite) such that:*

(a) *The set A is covered by the sequence, $A \subset \bigcup_{j \in \mathbb{N}} Q_j$.*
(b) *No point of $\mathbb{R}^n$ is in more than θ_n sets of the sequence $\{Q_j\}_{j \in \mathbb{N}}$.*
(c) *The sequence $\{Q_j\}_{j \in \mathbb{N}}$ can be divided into ξ_n families of disjoint sets.*

The numbers θ_n and ξ_n depend only on the dimension n.

Lemma 4.3.9 *Let $\varphi \in \Phi_w(\Omega)$ satisfy (A0), (A1) and (A2). Then there exist $\beta \in (0, 1)$ and $h \in L^1(\Omega) \cap L^\infty(\Omega)$ such that*

$$\varrho_\varphi(\beta \lambda \chi_{\{Mf > \lambda\}}) \leqslant \varrho_\varphi(f) + \int_{\{Mf > 2^{-n}\lambda\}} h(x)\, dx$$

for all $f \in L^\varphi(\Omega)$ with $\varrho_\varphi(f) \leqslant 1$, and for all $\lambda > 0$.

Proof In this proof we need the centered maximal operator, $M_\odot f(x)$, since the Besicovitch covering theorem requires centered balls. Fix $f \in L^\varphi(\Omega)$ with $\varrho_\varphi(f) \leqslant 1$ and $\lambda > 0$. Then $\{M_\odot f > \lambda\}$ is open, since $M_\odot$ is lower semicontinuous. Let K be an arbitrary compact subset of $\{M_\odot f > \lambda\}$. For every $x \in K$ there exists a ball B_x centered at x such that $\frac{1}{|B_x|} \int_{B_x \cap \Omega} |f|\, dy > \lambda$. We denote $\hat{B}_x := B_x \cap \Omega$ and observe that $\hat{B}_x \subset \{Mf > \lambda\}$ for the level-set of the non-centered maximal operator. From the family $\{B_x : x \in K\}$ we can select by the Besicovitch covering theorem (Theorem 4.3.8) a locally ξ_n-finite family $\mathcal{B}$, which covers K. The natural number ξ_n only depends on the dimension n. Let $m > 0$. By the key estimate (Theorem 4.3.2) with $p = 1$ there exists $\beta > 0$ such that

$$\varphi\left(x, \frac{\beta}{|B|} \int_{\hat{B}} |f| \, dy\right) \leqslant \frac{1}{|B|} \int_{\hat{B}} \varphi(y, |f|) + h(x) + h(y) \, dy$$

for all $B \in \mathcal{B}$. Now, (aInc)$_1$ with constant a, the inequality $\frac{1}{|B|} \int_{\hat{B}} |f| \, dy > \lambda$ for $B \in \mathcal{B}$, and the previous estimate imply that

$$\begin{aligned}
\varrho_\varphi\left(\frac{\beta}{2a\xi_n} \lambda \chi_K\right) &\leqslant \frac{1}{2\xi_n} \int_K \varphi(x, \beta\lambda) \, dx \\
&\leqslant \frac{1}{2\xi_n} \sum_{B \in \mathcal{B}} \int_{\hat{B}} \varphi\left(x, \frac{\beta}{|B|} \int_{\hat{B}} |f| \, dy\right) dx \\
&\leqslant \frac{1}{2\xi_n} \sum_{B \in \mathcal{B}} \frac{1}{|B|} \int_{\hat{B}} \left(\int_{\hat{B}} \varphi(y, |f|) + h(x) + h(y) \, dy\right) dx \\
&\leqslant \varrho_\varphi(f) + \int_{\{Mf > \lambda\}} h(y) \, dy.
\end{aligned}$$

Let $K_j \subset\subset \{M_\odot f > \lambda\}$ with $K_j \nearrow \{M_\odot f > \lambda\}$. Then monotone convergence in L^1 implies that

$$\varrho_\varphi\left(\frac{\beta}{2a\xi_n} \lambda \chi_{\{M_\odot f > \lambda\}}\right) \leqslant \varrho_\varphi(f) + \int_{\{Mf > \lambda\}} h(y) \, dy.$$

On the left-hand side we use $\{Mf > 2^n\lambda\} \subset \{M_\odot f > \lambda\}$ which follows from (4.3.1). The claim then follows by the change of variables $\lambda' := 2^n\lambda$. $\qquad \square$

We say that a operator $T : L^\varphi \to L^\varphi$ is of *weak type* if

$$\|\lambda \, \chi_{\{|Tf| > \lambda\}}\|_{L^\varphi(\Omega)} \lesssim \|f\|_{L^\varphi(\Omega)}$$

for all $f \in L^\varphi(\mathbb{R}^n)$ and all $\lambda > 0$. If we remove the assumption (aInc) from Theorem 4.3.4, we can still obtain a weak type estimate.

Theorem 4.3.10 (Weak Type Maximal Operator Estimate) *Let $\varphi \in \Phi_w(\Omega)$ satisfy (A0), (A1) and (A2). Then*

$$\|\lambda \chi_{\{Mf > \lambda\}}\|_\varphi \lesssim \|f\|_\varphi$$

for all $f \in L^\varphi(\Omega)$ and all $\lambda > 0$.

Proof Let $f \in L^\varphi(\mathbb{R}^n)$. We may assume that $\|f\|_\varphi > 0$ since otherwise $f = 0$ and there is nothing to prove. Let $\beta \in (0, 1)$ be the constant from Lemma 4.3.9. We define $c_1 := \|h\|_{L^1(\mathbb{R}^n)} + 1$ and $g := \beta f/\|f\|_\varphi$ and thus $\varrho_\varphi(g) \leqslant 1$ by the unit ball property. By (aInc)$_1$ and Lemma 4.3.9,

$$\varrho_\varphi\left(\frac{1}{ac_1}\beta\,\lambda\,\chi_{\{Mg>\lambda\}}\right) \leqslant \frac{1}{c_1}\varrho_\varphi(\beta\,\lambda\,\chi_{\{Mg>\lambda\}})$$

$$\leqslant \frac{1}{c_1}\left(\varrho_\varphi(g) + \int_{\{Mg>2^{-n}\lambda\}} h(x)\,dx\right)$$

$$\leqslant \frac{1}{c_1}\left(1 + \|h\|_{L^1(\mathbb{R}^n)}\right) \leqslant 1$$

and hence $\|\frac{1}{ac_1}\beta\,\lambda\,\chi_{\{Mg>\lambda\}}\|_\varphi \leqslant 1$ by the unit ball property (Lemma 3.2.3). Since $\{Mg>\lambda\} = \{Mf > \frac{\lambda}{\beta}\|f\|_\varphi\}$, we obtain

$$\frac{\beta}{\|f\|_\varphi}\left\|\frac{\lambda}{\beta}\|f\|_\varphi\,\chi_{\{Mf>\frac{\lambda}{\beta}\|f\|_\varphi\}}\right\|_\varphi = \|\lambda\,\chi_{\{Mg>\lambda\}}\|_\varphi \leqslant \frac{ac_1}{\beta}.$$

If we set $\lambda' := \|f\|_\varphi\lambda/\beta$, we find that $\|\lambda'\,\chi_{\{Mf>\lambda'\}}\|_\varphi \leqslant \frac{ac_1}{\beta^2}\|f\|_\varphi$. Since the this holds for all $\lambda' > 0$, we obtain the claim. $\qquad\square$

4.4 Averaging Operators and Applications

We now move on to averaging operators. Since extension of φ is non-trivial we pay particular attention to the role of the set $\Omega \subset \mathbb{R}^n$, in order to obtain results usable when our function is only defined on Ω.

Definition 4.4.1 A family $\mathcal{Q}$ of measurable sets $E \subset \mathbb{R}^n$ is called *locally N-finite*, $N \in \mathbb{N}$, if $\sum_{E \in \mathcal{Q}} \chi_E \leqslant N$ almost everywhere in $\mathbb{R}^n$. We simply say that $\mathcal{Q}$ is *locally finite* if it is locally N-finite for some $N \in \mathbb{N}$.

Note that a family $\mathcal{Q}$ of open, bounded sets $Q \subset \mathbb{R}^n$ is locally 1-finite if and only if the sets $Q \in \mathcal{Q}$ are pairwise disjoint.

Definition 4.4.2 For a family $\mathcal{B}$ of open, bounded sets $U \subset \mathbb{R}^n$ we define $T_\mathcal{B}\colon L^1_{\mathrm{loc}}(\Omega) \to L^0(\mathbb{R}^n)$ by

$$T_\mathcal{B}f(x) := \sum_{U \in \mathcal{B}} \frac{\chi_{U\cap\Omega}(x)}{|U|}\int_{U\cap\Omega}|f(y)|\,dy.$$

The operators $T_\mathcal{B}$ are called *averaging operators*.

Next we show that the averaging operator is bounded.

Theorem 4.4.3 (Averaging Operator Estimate) *Let* $\varphi \in \Phi_w(\Omega)$ *satisfy assumptions* (A0), (A1) *and* (A2). *Then the averaging operator*

$$T_{\mathcal{B}} : L^{\varphi}(\Omega) \to L^{\varphi}(\Omega)$$

is uniformly bounded for every locally N-finite family $\mathcal{B}$ of balls (or cubes).

Proof Let $f \in L^{\varphi}(\Omega)$ with $\|f\|_{\varphi} < 1$ and let $\mathcal{B}$ be a locally N-finite family of balls (or cubes). Then by the unit ball property (Lemma 3.2.3), $\varrho_{\varphi}(f) \leqslant 1$. Choose $\beta \in (0,1)$ and $h \in L^1(\mathbb{R}^n) \cap L^{\infty}(\mathbb{R}^n)$ as in the key estimate (Theorem 4.3.2) for $p = 1$. Denote $c_1 := 1 + 2\|h\|_{L^1(\mathbb{R}^n)}$ and $M_B f := \frac{1}{|B|} \int_{B \cap \Omega} |f| \, dy$. Then by (aInc)$_1$ and quasi-convexity (Corollary 2.2.2) with constant β_c,

$$\varrho_{\varphi}\left(\frac{\beta\beta_c}{ac_1 N} T_{\mathcal{B}} f\right) \leqslant \frac{1}{c_1} \varrho_{\varphi}\left(\frac{\beta\beta_c}{N} T_{\mathcal{B}} f\right) = \frac{1}{c_1} \int_{\Omega} \varphi\left(x, \frac{\beta\beta_c}{N} \sum_{B \in \mathcal{B}} \chi_{B \cap \Omega} M_B f\right) dx$$

$$\leqslant \frac{1}{c_1} \int_{\Omega} \sum_{B \in \mathcal{B}} \frac{\chi_{B \cap \Omega}(x)}{N} \varphi(x, \beta M_B f) \, dx$$

$$= \frac{1}{c_1 N} \sum_{B \in \mathcal{B}} \int_{B \cap \Omega} \varphi(x, \beta M_B f) \, dx.$$

By the key estimate (Theorem 4.3.2),

$$\int_{B \cap \Omega} \varphi(x, \beta M_B f) \, dx \leqslant \int_{B \cap \Omega} \left(\frac{1}{|B|} \int_{B \cap \Omega} \varphi(y, |f|) + h(x) + h(y) \, dy\right) dx$$

$$\leqslant \int_{B \cap \Omega} \varphi(y, |f|) \, dy + \int_{B \cap \Omega} h(x) \, dx + \int_{B \cap \Omega} h(y) \, dy.$$

Combining the two estimates, we find that

$$\varrho_{\varphi}\left(\frac{\beta\beta_c}{ac_1 N} T_{\mathcal{B}} f\right) \leqslant \frac{1}{1 + 2\|h\|_{L^1(\Omega)}} \left(\int_{\Omega} \varphi(y, |f|) \, dy + 2 \int_{\Omega} h(y) \, dy\right) \leqslant 1$$

where we used that $\mathcal{B}$ is locally N-finite. This implies by the unit ball property that $\|T_{\mathcal{B}} f\|_{\varphi} \leqslant \frac{ac_1 N}{\beta}$. A scaling argument yields $\|T_{\mathcal{B}} f\|_{\varphi} \leqslant \frac{ac_1 N}{\beta} \|f\|_{\varphi}$. □

In the rest of this section we prove a number of results under the assumption that $\varphi \in \Phi_w(\Omega)$ satisfies (A0), (A1) and (A2). In fact, all the results, except Theorem 4.4.7 and the latter claim in Proposition 4.4.11, hold under the weaker assumption that averaging operators T_Q are bounded from $L^{\varphi}(\Omega)$ to $L^{\varphi}(\Omega)$ uniformly for all locally 1-finite families Q of cubes in Ω. This assumption is called *class $\mathcal{A}$* in [34].

Let $f \in L^1_{\mathrm{loc}}(\Omega)$, $g \in L^1_{\mathrm{loc}}(\mathbb{R}^n)$ and $x \in \Omega$. Let us write

$$f * g(x) := \int_\Omega f(y)g(x - y)\,dy.$$

Lemma 4.4.4 *Let* $\varphi \in \Phi_{\mathrm{w}}(\Omega)$ *satisfy* (A0), (A1) *and* (A2). *Then*

$$\left\| f * \frac{\chi_B}{|B|} \right\|_\varphi \leqslant \left\| |f| * \frac{\chi_B}{|B|} \right\|_\varphi \lesssim \|f\|_\varphi$$

for all $f \in L^\varphi(\Omega)$ *and all balls (or cubes)* $B \subset \mathbb{R}^n$ *centered at* 0.

Proof Let $B \subset \mathbb{R}^n$ be a ball with center at 0 and $f \in L^\varphi(\Omega)$. For $k \in \mathbb{Z}^n$ let $B_k := \mathrm{diam}(B)k + B$, be the translation of B by the vector $\mathrm{diam}(B)k$. Then the balls $\{B_k\}_k$ are disjoint. Moreover, we can split the set $\{3B_k\}_k$ into 3^n locally 1-finite families $\mathcal{B}_j$, $j = 1, \dots, 3^n$. For every $k \in \mathbb{Z}^n$ and every $x \in B_k \cap \Omega$ we note that $\chi_B(x - y) = 0$ unless $y \in 3B_k$ since B has center at the origin. Hence

$$f * \frac{\chi_B}{|B|}(x) \leqslant |f| * \frac{\chi_B}{|B|}(x) \leqslant \frac{3^n}{|3B|} \int_{3B_k \cap \Omega} |f(y)|\,dy \leqslant 3^n \sum_{j=1}^{3^n} T_{\mathcal{B}_j} f(x)$$

and we conclude, by Theorem 4.4.3, that

$$\left\| f * \frac{\chi_B}{|B|} \right\|_\varphi \leqslant 3^n \sum_{j=1}^{3^n} \left\| T_{\mathcal{B}_j} f \right\|_\varphi \lesssim \|f\|_\varphi. \qquad \square$$

With this lemma we prove convolution estimates for bell shaped functions.

Definition 4.4.5 A non-negative function $\sigma \in L^1(\mathbb{R}^n)$ is called *bell shaped* if it is radially decreasing and radially symmetric. The function

$$\Sigma(x) := \sup_{y \notin B(0, |x|)} |\sigma(y)|$$

is called the *least bell shaped majorant of* σ.

Recall that we defined the L^1-scaling $\sigma_\varepsilon(x) := \frac{1}{\varepsilon^n} \sigma(\frac{x}{\varepsilon})$, for $\varepsilon > 0$. Then, by a change of variables, $\|\sigma_\varepsilon\|_1 = \|\sigma\|_1$. Recall also that if $B := B(x, r)$, then $\varepsilon B = B(x, \varepsilon r)$ is the scaled ball with the same center.

Lemma 4.4.6 *Let* $\varphi \in \Phi_{\mathrm{w}}(\Omega)$ *satisfy* (A0), (A1) *and* (A2). *Let* $\sigma \in L^1(\mathbb{R}^n)$, $\sigma \geqslant 0$, *have integrable least bell shaped majorant* Σ. *Then*

$$\|f * \sigma_\varepsilon\|_\varphi \lesssim \|\Sigma\|_1 \|f\|_\varphi$$

for all $f \in L^\varphi(\Omega)$. *Moreover,* $|f * \sigma_\varepsilon| \leqslant 2 \|\Sigma\|_1 Mf$ *for all* $f \in L^1_{\mathrm{loc}}(\Omega)$.

Proof We may assume without loss of generality that $f \geqslant 0$. Since $\sigma \leqslant \Sigma$, we may further assume that σ is already bell shaped. Any bell shaped function σ can be approximated from above by functions of type

$$g := \sum_{k=1}^{\infty} a_k \frac{\chi_{B_k}}{|B_k|},$$

where $a_k \in [0, \infty)$, B_k are balls with center at 0 and

$$\|g\|_{L^1(\mathbb{R}^n)} = \sum_{k=1}^{\infty} a_k \leqslant 2 \, \|\sigma\|_{L^1(\mathbb{R}^n)}.$$

Then $0 \leqslant \sigma_\varepsilon \leqslant g_\varepsilon$ and

$$g_\varepsilon = \sum_{k=1}^{\infty} a_k \frac{\chi_{\varepsilon B_k}}{|\varepsilon B_k|}.$$

By Theorem 4.4.3, the averaging operator is bounded. Using Lemma 4.4.4 and the quasi-triangle inequality (Corollary 3.2.5), we estimate

$$\left\| f * \sigma_\varepsilon \right\|_\varphi \leqslant \left\| f * g_\varepsilon \right\|_\varphi = \left\| \sum_{k=1}^{\infty} f * \left(a_k \frac{\chi_{\varepsilon B_k}}{|\varepsilon B_k|} \right) \right\|_\varphi$$

$$\leqslant Q \sum_{k=1}^{\infty} a_k \left\| f * \frac{\chi_{\varepsilon B_k}}{|\varepsilon B_k|} \right\|_\varphi \lesssim \sum_{k=1}^{\infty} a_k \, \|f\|_\varphi$$

$$\lesssim \|\sigma\|_1 \|f\|_\varphi.$$

Analogously, for $f \in L^1_{\mathrm{loc}}(\mathbb{R}^n)$ we estimate point-wise

$$|f * \sigma_\varepsilon| \leqslant \sum_{k=1}^{\infty} |f| * \left(a_k \frac{\chi_{\varepsilon B_k}}{|\varepsilon B_k|} \right) \leqslant \sum_{k=1}^{\infty} a_k M f \leqslant 2 \, \|\sigma\|_1 M f. \qquad \square$$

Theorem 4.4.7 (Mollification) *Let* $\varphi \in \Phi_{\mathrm{w}}(\Omega)$ *satisfy* (A0). *Let* $\sigma \in L^1(\mathbb{R}^n)$, $\sigma \geqslant 0$, *with* $\|\sigma\|_1 = 1$ *have integrable least bell shaped majorant* Σ. *Then* $f * \sigma_\varepsilon \to f$ *almost everywhere as* $\varepsilon \to 0^+$ *for* $f \in L^\varphi(\Omega)$. *If additionally* φ *satisfies* (aDec), (A1) *and* (A2), *then* $f * \sigma_\varepsilon \to f$ *in* $L^\varphi(\Omega)$.

Proof Let $f \in L^\varphi(\Omega)$ with $\|f\|_\varphi \leqslant 1$. By Lemma 3.7.7 we can split f into $f = f_1 + f_2$ with $f_1 \in L^1(\Omega)$ and $f_2 \in L^\infty(\Omega)$. From [124, Theorem 2, p. 62] we deduce $f_j * \sigma_\varepsilon \to f_j$ almost everywhere, $j = 0, 1$. By linearity of convolution, $f * \sigma_\varepsilon \to f$ almost everywhere.

Let us then assume that φ satisfies (aDec)$_q$, (A1) and (A2). By Proposition 3.5.1, simple functions are dense in $L^\varphi(\Omega)$. Let $\delta > 0$ be arbitrary and let g be a simple function with $\|f - g\|_\varphi \leqslant \delta$. Then

$$\|f * \sigma_\varepsilon - f\|_\varphi \leqslant \|g * \sigma_\varepsilon - g\|_\varphi + \|(f - g) * \sigma_\varepsilon - (f - g)\|_\varphi =: (I) + (II).$$

Since g is a simple function, we have $g \in L^1(\Omega) \cap L^q(\Omega)$. Thus the classical theorem on mollification, see [124, Theorem 2, p. 62] again, implies that $g * \sigma_\varepsilon \to g$ in $L^1(\Omega) \cap L^q(\Omega)$. By Lemma 3.7.7, $g * \sigma_\varepsilon \to g$ in $L^\varphi(\Omega)$. This proves $(I) \to 0$ for $\varepsilon \to 0^+$. On the other hand, Lemma 4.4.6 implies that

$$(II) = \|(f - g) * \sigma_\varepsilon - (f - g)\|_\varphi \lesssim \|(f - g) * \sigma_\varepsilon\|_\varphi + \|f - g\|_\varphi$$

$$\lesssim \|f - g\|_\varphi \leqslant \delta.$$

Combining the estimates for (I) and (II), we conclude that

$$\limsup_{\varepsilon \to 0^+} \|f * \sigma_\varepsilon - f\|_\varphi \lesssim \delta.$$

Since $\delta > 0$ was arbitrary, this yields $\|f * \sigma_\varepsilon - f\|_\varphi \to 0$ as $\varepsilon \to 0^+$.   □

If $\varphi(x, t) = t^p$, then a short calculation gives $\|\chi_D\|_p = |D|^{1/p}$ for every open set D. Unfortunately in the generalized Orlicz case, it is complicated to calculate the norm of a characteristic function. The next lemma shows that there is nevertheless a simple connection between $\|\chi_D\|_\varphi$ and $\|\chi_D\|_{\varphi^*}$ provided that D is regular enough. Furthermore, it will be used in Proposition 4.4.11 to calculate the norm of characteristic functions of balls.

Lemma 4.4.8 *Let $\varphi \in \Phi_w(\Omega)$ satisfy (A0), (A1) and (A2). Let $b \geqslant 1$ be a constant. If $D \subset \Omega$ and there exist balls B and B' such that $B \subset D \subset B'$ and $|B'| \leqslant b|B|$, then*

$$\|\chi_D\|_\varphi \|\chi_D\|_{\varphi^*} \approx |D|,$$

where the implicit constant depends on b but not D.

Proof By Theorem 4.4.3, averaging operators are bounded on $L^\varphi(\Omega)$. By Proposition 2.4.5, $\varphi \simeq \varphi^{**}$ and thus averaging operators are bounded on $L^{\varphi^{**}}$. Thus by the norm conjugate formula of L^φ (Theorem 3.4.6) we obtain

$$\|\chi_D\|_\varphi \|\chi_D\|_{\varphi^*} \lesssim \|\chi_D\|_\varphi \sup_{\|g\|_{\varphi^{**}} \leqslant 1} \int_D g\, dx = |D| \sup_{\|g\|_{\varphi^{**}} \leqslant 1} \left\| \chi_D \fint_D g\, dx \right\|_\varphi$$

$$\leqslant |D| \sup_{\|g\|_{\varphi^{**}} \leqslant 1} \left\| \frac{1}{|B|} \int_{B' \cap \Omega} g\, dx \right\|_\varphi$$

$$\leqslant b|D| \sup_{\|g\|_{\varphi^{**}} \leqslant 1} \left\| \frac{1}{|B'|} \int_{B' \cap \Omega} g \, dx \right\|_{\varphi^{**}}$$

$$\leqslant |D| \sup_{\|g\|_{\varphi^{**}} \leqslant 1} A \|g\|_{\varphi^{**}} = A|D|,$$

where A is the embedding constant of the averaging operator. The other direction follows from Hölder's inequality (Lemma 3.2.11). $\qquad\square$

The previous result shows that the boundedness of averaging operators implies $\|\chi_D\|_\varphi \|\chi_D\|_{\varphi^*} \approx |D|$. The next result is a partial converse, in that this condition is shown to imply the boundedness of certain averaging operators.

Lemma 4.4.9 *Let $\varphi \in \Phi_w(\Omega)$ and assume that $\|\chi_D\|_\varphi \|\chi_D\|_{\varphi^*} \approx |D|$ for a bounded open set $D \subset \Omega$. Then the averaging operator $T_{\{D\}}$ is bounded from $L^\varphi(\Omega)$ to $L^\varphi(\Omega)$ and the embedding constant depends only on the implicit constant.*

Proof Using Hölder's inequality (Lemma 3.2.11) we get, for all $f \in L^\varphi(\mathbb{R}^n)$,

$$\|T_{\{D\}} f\|_\varphi = \left\| \chi_D \fint_D f \, dy \right\|_\varphi = \|\chi_D\|_\varphi \frac{1}{|D|} \int_\Omega \chi_D(y)|f(y)| \, dy$$

$$\leqslant \|\chi_D\|_\varphi \frac{2}{|D|} \|\chi_D\|_{\varphi^*} \|f\|_\varphi.$$

Then $\|\chi_D\|_\varphi \|\chi_D\|_{\varphi^*} \approx |D|$ yields the boundedness of $T_{\{D\}}$. $\qquad\square$

The next lemma is used to estimate norms of characteristic functions of balls.

Lemma 4.4.10 *Let $\varphi \in \Phi_w(\Omega)$ and let $E \subset \Omega$ be measurable with $|E| \in (0, \infty)$. Then for all $t \geqslant 0$*

$$t \lesssim \fint_E \varphi^{-1}(x, t) \, dx \fint_E (\varphi^*)^{-1}(x, t) \, dx.$$

Proof It suffices to consider the case $t > 0$. By Theorem 2.4.8,

$$\varphi^{-1}(x, t) \gtrsim \frac{t}{(\varphi^*)^{-1}(x, t)}$$

for all $t > 0$ and almost all $x \in E$. By Jensen's inequality for $z \mapsto \frac{1}{z}$,

$$\fint_E \varphi^{-1}(x, t) \, dx \gtrsim t \fint_E \frac{1}{(\varphi^*)^{-1}(x, t)} \, dx \geqslant \frac{t}{\fint_E (\varphi^*)^{-1}(x, t) \, dx}. \qquad\square$$

Proposition 4.4.11 *Let $\varphi \in \Phi_w(\Omega)$ satisfy (A0), (A1) and (A2). Then for every ball $B \subset \Omega$ we have*

$$\|\chi_B\|_\varphi \approx \frac{1}{\fint_B \varphi^{-1}(x, \frac{1}{|B|}) \, dx}.$$

If furthermore $|B| \leqslant 1$, then for every $y \in B$

$$\|\chi_B\|_\varphi \approx \frac{1}{\varphi^{-1}(y, \frac{1}{|B|})}.$$

Proof Let $B \subset \mathbb{R}^n$ be a ball. We first prove

$$1 \lesssim |B| \fint_B (\varphi^*)^{-1}\left(x, \tfrac{1}{|B|}\right) dx \fint_B \varphi^{-1}\left(x, \tfrac{1}{|B|}\right) dx \qquad (4.4.1)$$
$$\leqslant 2\|\chi_B\|_\varphi \fint_B \varphi^{-1}\left(x, \tfrac{1}{|B|}\right) dx \leqslant 2A,$$

where A is from Theorem 4.4.3.

Let us define

$$f(x) := \chi_B(x)\,\varphi^{-1}\left(x, \frac{1}{|B|}\right) \quad \text{and} \quad g(x) := \chi_B(x)\,(\varphi^*)^{-1}\left(x, \frac{1}{|B|}\right),$$

and note that $\varrho_\varphi(\frac{f}{\lambda}) \leqslant 1$ and $\varrho_{\varphi^*}(\frac{g}{\lambda}) \leqslant 1$ for every $\lambda > 1$, by Lemma 2.3.9(a). Thus, $\|f\|_\varphi \leqslant 1$ and $\|g\|_{\varphi^*} \leqslant 1$ by the unit ball property. The first inequality in (4.4.1) follows directly from Lemma 4.4.10. The second one follows from Hölder's inequality (Lemma 3.2.11):

$$|B| \fint_B (\varphi^*)^{-1}\left(x, \tfrac{1}{|B|}\right) dx = \int_B g\,dx \leqslant 2\|\chi_B\|_\varphi \|g\|_{\varphi^*} \leqslant 2\|\chi_B\|_\varphi.$$

The third inequality follows since the averaging operator is bounded (Theorem 4.4.3):

$$\|\chi_B\|_\varphi \fint_B \varphi^{-1}\left(x, \tfrac{1}{|B|}\right) dx = \|\chi_B\|_\varphi \fint_B f\,dx = \|T_{\{B\}} f\|_\varphi \leqslant A\|f\|_\varphi \leqslant A.$$

So we have proved (4.4.1).

Since $\int_B \varphi^{-1}\left(x, \tfrac{1}{|B|}\right) dx > 0$, we obtain from the first and third inequalities of (4.4.1) that

$$\frac{1}{2 \fint_B \varphi^{-1}\left(x, \tfrac{1}{|B|}\right) dx} \lesssim \|\chi_B\|_\varphi \leqslant \frac{A}{\fint_B \varphi^{-1}\left(x, \tfrac{1}{|B|}\right) dx}.$$

This is the first claim of the proposition.

Assume next that $|B| \leqslant 1$. Then it follows from (A1) that $\varphi^{-1}(y, \frac{1}{|B|}) \approx \varphi^{-1}(x, \frac{1}{|B|})$ for every $x, y \in B$. Thus

$$\fint_B \varphi^{-1}\left(x, \tfrac{1}{|B|}\right) dx \approx \fint_B \varphi^{-1}\left(y, \tfrac{1}{|B|}\right) dx = \varphi^{-1}\left(y, \tfrac{1}{|B|}\right),$$

which implies the second claim of the proposition. $\square$

Chapter 5
Extrapolation and Interpolation

In this chapter, we develop two techniques which allow us to transfer results of harmonic analysis from one setting to another: extrapolation and interpolation. This section is based on the paper [29]; similar extrapolation results were derived earlier by Maeda, Sawano and Shimomura [87] with the stronger assumptions from Sect. 7.3.

In Sect. 5.1, we first give some necessary definitions about weights, and we introduce the abstract formalism of families of extrapolation pairs. In Sect. 5.2 we study some properties of Φ-functions relevant for extrapolation. Section 5.3 contains the main results. We then deduce a number of immediate corollaries that combine the extrapolation result with the sufficient conditions (A0), (A1) and (A2) from previous chapters. In Sect. 5.4 we give applications of extrapolation to

- the maximal operator,
- Calderón–Zygmund singular integrals, and
- the Riesz potential and fractional maximal operators.

This list shows the versatility of extrapolation in proving inequalities with very little additional work. However, there is a small price to pay: we need to assume (aDec), since the technique uses the maximal operator in the dual space. Also, in contrast to variable exponent spaces, the question of extension is non-trivial, so unless one has weighted results in domains, extrapolation gives only results in the whole space $\mathbb{R}^n$. Consequently, it is of interest also to find direct proofs for the above mentioned operators and results proved based on them, cf. e.g. Corollary 6.3.4. Some results exist for other operators, as well, for instance the maximal operator with rough kernel by Rafeiro and Samko [116], the sharp maximal operator has been studied by Yang, Yang and Yuan in [127], and the Hardy operator by Karaman [69].

Interpolation has not proved to be as useful in variable exponent spaces. The reason is that there are no special variable exponent spaces where proving results would be easier (like L^2, L^1 or L^∞ for constant exponents). Furthermore, variable exponent spaces cannot be obtained by interpolating constant exponent spaces. The

© Springer Nature Switzerland AG 2019
P. Harjulehto, P. Hästö, *Orlicz Spaces and Generalized Orlicz Spaces*,
Lecture Notes in Mathematics 2236,
https://doi.org/10.1007/978-3-030-15100-3_5

same caveats hold for generalized Orlicz spaces. However, we are not aware of any method to obtain compact embeddings directly by extrapolation. Therefore, we also study complex interpolation, in Sect. 5.5, which combined with the Sobolev embedding theorem allows us to extend the Rellich–Kondratchov Theorem to generalized Orlicz spaces in Sect. 6.3.

5.1 Weights and Classical Extrapolation

Extrapolation was introduced by Rubio de Francia [120]. It is a powerful tool in the study of weighted norm inequalities: roughly, it shows that if an operator T is bounded on the weighted spaces $L^2(w)$ for every weight in the Muckenhoupt class A_2, then for all $p \in (1, \infty)$ the operator T is bounded on $L^p(w)$ when $w \in A_p$. It can be used to prove norm inequalities on Banach spaces, provided that the maximal operator is bounded on the dual space (or, more precisely, the associate space). This approach was first used for non-standard growth in [28], where norm inequalities in variable exponent Lebesgue spaces were proved by extrapolation, see also [27, 30, 31].

We give some preliminary definitions and results about weights and the classical theory of Rubio de Francia extrapolation, as well as more recent generalizations. For more information and proofs, we refer the reader to [30, 37] and the references therein.

By a *weight* we mean a positive, locally integrable function w. For $p \in [1, \infty)$, the weighted Lebesgue space $L^p(\mathbb{R}^n, w)$ consists of all $f \in L^0$ such that

$$\|f\|_{L^p(\mathbb{R}^n, w)} := \left(\int_{\mathbb{R}^n} |f|^p w \, dx \right)^{1/p} < \infty.$$

For $p \in (1, \infty)$, a weight w is in the *Muckenhoupt class A_p*, denoted by $w \in A_p$, if

$$[w]_{A_p} := \sup_Q \left(\fint_Q w \, dx \right) \left(\fint_Q w^{1-p'} \, dx \right)^{p-1} < \infty,$$

where the supremum is taken over all cubes with sides parallel to the coordinate axes, and $\frac{1}{p} + \frac{1}{p'} = 1$. (Equivalently, we can replace cubes by balls.) When $p = 1$, we say $w \in A_1$ if

$$[w]_{A_1} := \sup_Q \left(\fint_Q w \, dx \right) \operatorname*{ess\,sup}_{x \in Q} \frac{1}{w(x)} < \infty.$$

Given $1 < p < q < \infty$, $A_1 \subsetneq A_p \subsetneq A_q$. A simple example of a Muckenhoupt A_p-weight is $w(x) = |x|^\alpha$ with $\alpha \in (-n, n(p-1))$.

We follow the approach of [30] and consider inequalities of so-called extrapolation pairs (f, g) rather than inequalities relating the norms of Tf and f. We will consider families of pairs of non-negative measurable functions,

$$\mathcal{F} := \{(f, g)\},$$

with implicit additional restrictions on f and g. In our extrapolation theorems we will assume that we have weighted norm inequalities

$$\|f\|_{L^p(\Omega, w)} \leqslant C([w]_{A_q})\|g\|_{L^p(\Omega, w)} \tag{5.1.1}$$

and use them to deduce generalized Orlicz space inequalities

$$\|f\|_{L^\varphi(\Omega)} \lesssim \|g\|_{L^\varphi(\Omega)}. \tag{5.1.2}$$

More precisely, in order to prove (5.1.2) for a fixed pair (f, g), we need (5.1.1) for a weight $w \in A_q$ that we construct in the course of the proof. The problem is that in this abstract setting we cannot rule out *a priori* that $\|f\|_{L^p(\Omega, w)} = \infty$ for the constructed weigth.

To avoid this problem we adopt the following convention. Given a family $\mathcal{F}$ of extrapolation pairs, if we write

$$\|f\|_{L^p(\Omega, w)} \leqslant C([w]_{A_p})\|g\|_{L^p(\Omega, w)}, \qquad (f, g) \in \mathcal{F},$$

then we mean this inequality holds for a given weight $w \in A_p$ for all pairs (f, g) such that the left-hand side is finite. The same convention applies also for generalized Orlicz spaces in place of $L^p(\Omega, w)$.

We conclude the section by stating the classical extrapolation result of Rubio de Francia. For a proof, see [30, Theorem 3.9, Corollary 3.12].

Theorem 5.1.1 *Given a family of extrapolation pairs $\mathcal{F}$, suppose that for some $p_0 \in [1, \infty)$ and every $w_0 \in A_{p_0}$,*

$$\|f\|_{L^{p_0}(\Omega, w_0)} \leqslant C(n, p_0, [w_0]_{A_{p_0}})\|g\|_{L^{p_0}(\Omega, w_0)}, \qquad (f, g) \in \mathcal{F}.$$

Then for every $p \in (1, \infty)$, and every $w \in A_p$,

$$\|f\|_{L^p(\Omega, w)} \leqslant C(n, p, [w]_{A_p})\|g\|_{L^p(\Omega, w)}, \qquad (f, g) \in \mathcal{F}.$$

Moreover, for every $p, q \in (1, \infty)$, $w \in A_p$ and pairs $(f_k, g_k) \subset \mathcal{F}$,

$$\left\|\left(\sum_k |f_k|^q\right)^{1/q}\right\|_{L^p(\Omega, w)} \leqslant C(n, p, q, [w]_{A_p})\left\|\left(\sum_k |g_k|^q\right)^{1/q}\right\|_{L^p(\Omega, w)}.$$

5.2 Rescaling and Conditions (A0), (A1) and (A2)

In this section we collected tools needed to work with our assumptions in the context of extrapolation. Suppose that $\varphi \in \Phi_w(\Omega)$ satisfies (aInc)$_p$. Then

$$\varphi_p(x, t) := \varphi\left(x, t^{\frac{1}{p}}\right)$$

is also a weak Φ-function and it follows directly from the definition of the norm that

$$\left\| |v|^p \right\|_{\varphi_p} = \|v\|_\varphi^p.$$

This identity will be referred to as *rescaling*. The next lemma describes how rescaling behaves under conjugation. Note that there φ satisfies (aInc)$_p$ if and only if ψ satisfies (aInc)$_q$, by Proposition 2.3.7. We write $\varphi_p^* := (\varphi_p)^*$.

Lemma 5.2.1 *Let $1 \leqslant p \leqslant q < \infty$, $\gamma := \frac{1}{p} - \frac{1}{q}$ and $\varphi, \psi \in \Phi_w(\Omega)$ be such that $\varphi^{-1}(x, t) \approx t^\gamma \psi^{-1}(x, t)$ and φ satisfies (aInc)$_p$. Then*

$$\left\| |v|^{\frac{p}{q}} \right\|_{\varphi_p^*} \approx \|v\|_{\psi_q^*}^{\frac{p}{q}}.$$

Proof By Theorem 2.4.8, $\varphi_p^{-1}(x, t)(\varphi_p^*)^{-1}(x, t) \approx t$ and from the definition, $\varphi_p^{-1}(x, t) = \varphi^{-1}(x, t)^p$. Hence, we have that $(\varphi_p^*)^{-1}(x, t) \approx t\left(\varphi^{-1}(x, t)\right)^{-p}$. Analogously, $\psi^{-1}(x, t) \approx (t/(\psi_q^*)^{-1}(x, t))^{1/q}$. By assumption, $\varphi^{-1}(x, t) \approx t^\gamma \psi^{-1}(x, t)$. Therefore,

$$(\varphi_p^*)^{-1}(x, t) \approx t\left(\varphi^{-1}(x, t)\right)^{-p} \approx t^{1-p\gamma}\left(\psi^{-1}(x, t)\right)^{-p}$$

$$\approx t^{1-p\gamma-\frac{p}{q}}\left((\psi_q^*)^{-1}(x, t)\right)^{\frac{p}{q}}.$$

Since $1 - p\gamma - \frac{p}{q} = 0$, we have shown by Theorem 2.3.6 that $\varphi_p^*(x, t) \simeq \psi_q^*(x, t^{q/p})$, and so the claim follows by the rescaling identity. $\square$

To describe the impact of rescaling we define an operator on Φ-functions: for $\varphi \in \Phi_w$ and $\alpha > 0$, let $T_\alpha(\varphi)(x, t) := \varphi(x, t^{\frac{1}{\alpha}})$. We study the behavior of our conditions under this transformation.

Proposition 5.2.2 *Let $\varphi \in \Phi_w(\Omega)$. If φ satisfies (A0), (A1) or (A2), then so does $T_\alpha(\varphi)$. Furthermore, T_α maps (aInc)$_p$ to (aInc)$_{p/\alpha}$ and (aDec)$_q$ to (aDec)$_{q/\alpha}$.*

Proof By choosing $t = 1$ in $(T_\alpha\varphi)^{-1}(x, t) = \varphi^{-1}(x, t)^\alpha$, we see that (A0) is invariant under T_α. From the same identity, we see that (A1) and (A2) are invariant. We observe that φ satisfies (aInc)$_p$ if and only if $\frac{\varphi(x, t^{1/p})}{t}$ is almost increasing.

Since $T_\alpha \varphi(x, t^{\frac{\alpha}{p}}) = \varphi(x, t^{\frac{1}{p}})$, $T_\alpha \varphi$ then satisfies $(\text{aInc})_{p/\alpha}$. The case of (aDec) is analogous.   □

In off-diagonal extrapolation, we have two Φ-functions, φ and ψ. The following lemma clarifies how one can be constructed from the other.

Lemma 5.2.3 *Let* $1 \leqslant p \leqslant q < \infty$, $\gamma := \frac{1}{p} - \frac{1}{q}$ *and let* $\varphi \in \Phi_w(\Omega)$ *satisfy* $(\text{aInc})_p$ *and* $(\text{aDec})_{1/\gamma}$. *Then there exists* $\psi \in \Phi_w(\Omega)$ *satisfying* $(\text{aInc})_q$ *such that* $t^{-\gamma} \varphi^{-1}(x, t) \approx \psi^{-1}(x, t)$. *Moreover, if* φ *satisfies* $(\text{aDec})_{1/r}$ *for* $r > \gamma$, *then* ψ *can be chosen to satisfy* $(\text{aDec})_{1/(r-\gamma)}$.

Proof Since φ satisfies $(\text{aDec})_{1/\gamma}$ and $(\text{aInc})_p$, it follows by Proposition 2.3.7 that φ^{-1} satisfies $(\text{aInc})_\gamma$ and $(\text{aDec})_{1/p}$. Let us set $f(x, 0) := 0$ and

$$f(x, s) := \sup_{t \in (0,s]} t^{-\gamma} \varphi^{-1}(x, t) = \sup_{t \in (0,s] \cap \mathbb{Q}} t^{-\gamma} \varphi^{-1}(x, t);$$

the equality follows since φ^{-1} is left-continuous. We show that $f \in \Phi_w^{-1}(\Omega)$. By $(\text{aInc})_\gamma$,

$$s^{-\gamma} \varphi^{-1}(x, s) \leqslant f(x, s) \leqslant a s^{-\gamma} \varphi^{-1}(x, s)$$

so that $f(x, s) \approx s^{-\gamma} \varphi^{-1}(x, s)$. Clearly f is increasing, $f(x, t) = 0$ if and only if $t = 0$, and $f(x, t) = \infty$ if and only if $t = \infty$. Using $f(x, s) \approx s^{-\gamma} \varphi^{-1}(x, s)$ and $(\text{aDec})_{1/p}$ of φ^{-1}, we obtain for $t < s$ that

$$\frac{f(x, s)}{s^{\frac{1}{q}}} \leqslant \frac{a s^{-\gamma} \varphi^{-1}(x, s)}{s^{\frac{1}{q}}} = a \frac{\varphi^{-1}(x, s)}{s^{\frac{1}{p}}} \leqslant a^2 \frac{\varphi^{-1}(x, t)}{t^{\frac{1}{p}}} \leqslant a^2 \frac{f(x, t)}{t^{\frac{1}{q}}}$$

and thus f satisfies $(\text{aDec})_{1/q}$ and hence also $(\text{aDec})_1$. To show that f is left-continuous, let $s > 0$, $s_j \to s^-$ and $(s_j) \subset [0, \infty)$ and $\varepsilon > 0$. Let $t_0 \in [0, s]$ be such that $t_0^{-\gamma} \varphi^{-1}(x, t_0) \geqslant f(x, s) - \varepsilon$. Now $t_j := \min\{s_j, t_0\} \to t_0^-$ and hence by the left-continuity of φ^{-1} we obtain that

$$\lim_{j \to \infty} f(x, s_j) \geqslant \lim_{j \to \infty} t_j^{-\gamma} \varphi^{-1}(x, t_j) = t_0^{-\gamma} \varphi^{-1}(x, t_0) \geqslant f(x, s) - \varepsilon.$$

Letting $\varepsilon \to 0^+$ we obtain $\lim_{j \to \infty} f(x, s_j) \geqslant f(x, s)$. The opposite inequality follows since f is increasing and so f is left-continuous. By Lemma 2.5.12, $x \mapsto \varphi^{-1}(x, t)$ is measurable and hence as a countable supremum of measurable functions $x \mapsto f(x, t)$ is measurable for every $t \geqslant 0$. Thus $f \in \Phi_w^{-1}(\Omega)$ and hence by Proposition 2.5.14 we have $\psi := f^{-1} \in \Phi_w(\Omega)$. Since f satisfies $(\text{aDec})_{1/q}$ it follows by Proposition 2.3.7 that ψ satisfies $(\text{aInc})_q$.

If φ satisfies $(aDec)_{1/r}$, then φ^{-1} satisfies $(aInc)_r$ by Proposition 2.3.7 and hence

$$\frac{f(x,t)}{t^{r-\gamma}} \leqslant \frac{a\varphi^{-1}(x,t)}{t^r} \leqslant a^2 \frac{\varphi^{-1}(x,s)}{s^r} \leqslant a^2 \frac{f(x,s)}{s^{r-\gamma}}$$

for $t < s$, i.e. f satisfies $(aInc)_{r-\gamma}$. Thus $\psi = f^{-1}$ satisfies $(aDec)_{1/(r-\gamma)}$. □

Remark 5.2.4 Notice that if $t^{-\gamma}\varphi^{-1}(x,t) \approx \psi^{-1}(x,t)$ and $\psi \in \Phi_w(\Omega)$, then φ satisfies $(aDec)_{1/\gamma}$, so this assumption in the previous result is also necessary.

5.3 Diagonal and Off-Diagonal Extrapolation

In this section, we prove Theorem 5.3.2 as a consequence of the following more general result expressed in terms of extrapolation pairs. Notice that we consider weights in $\mathbb{R}^n$ but function spaces on Ω.

We generalize the off-diagonal extrapolation theorem of Harboure, Macias and Segovia [51]. In variable exponent spaces, this result was proved in [28].

Theorem 5.3.1 *Let* $1 \leqslant p \leqslant q < \infty$, $\gamma := \frac{1}{p} - \frac{1}{q}$ *and* $\varphi, \psi \in \Phi_w(\Omega)$ *be such that* $\varphi^{-1}(x,t) \approx t^\gamma \psi^{-1}(x,t)$ *and* φ *satisfies* $(aInc)_p$. *Given a family of extrapolation pairs* $\mathcal{F}$, *suppose that*

$$\|f\|_{L^q(\Omega,w)} \leqslant C([w]_{A_1})\|g\|_{L^p(\Omega,w^{p/q})}, \qquad (f,g) \in \mathcal{F}, \qquad (5.3.1)$$

for all $w \in A_1$. *If the maximal operator is bounded on* $L^{\psi_q^*}(\Omega)$, *then*

$$\|f\|_{L^\psi(\Omega)} \lesssim \|g\|_{L^\varphi(\Omega)}, \qquad (f,g) \in \mathcal{F}.$$

Proof We begin the proof by using the Rubio de Francia iteration algorithm. Let M be the uncentered maximal operator over cubes with sides parallel to the coordinate axes and $m := \|M\|_{L^{\psi_q^*} \to L^{\psi_q^*}}$ be the operator-norm. Define $\mathcal{R} : L^0(\mathbb{R}^n) \to [0, \infty]$ by

$$\mathcal{R}h(x) := \sum_{k=0}^{\infty} \frac{M^k h(x)}{2^k m^k},$$

where for $k \geqslant 1$, M^k denotes k iterations of the maximal operator, and $M^0 h := |h|$. We show the following properties for all $h \in L^{\psi_q^*}(\Omega)$:

(A). $|h| \leqslant \mathcal{R}h$,
(B). $\|\mathcal{R}h\|_{L^{\psi_q^*}(\Omega)} \leqslant 2Q\|h\|_{L^{\psi_q^*}(\Omega)}$,
(C). $\mathcal{R}h \in A_1$ and $[\mathcal{R}h]_{A_1} \leqslant 2m$.

Property (A) holds since $\mathcal{R}h \geqslant M^0 h = |h|$. Property (B) holds by the quasi-triangle inequality (Corollary 3.2.5), since

$$\left\| \frac{M^k h}{2^k m^k} \right\|_{L^{\psi_q^*}(\Omega)} = \frac{\|M^k h\|_{L^{\psi_q^*}(\Omega)}}{2^k m^k} \leqslant \frac{m \|M^{k-1} h\|_{L^{\psi_q^*}(\Omega)}}{2^k m^k} \leqslant \cdots \leqslant \frac{\|h\|_{L^{\psi_q^*}(\Omega)}}{2^k}.$$

Let us then prove (C). We may assume that $h \neq 0$, since the claim is trivial otherwise. By the definition of $\mathcal{R}$ and the sublinearity of the maximal operator, we obtain

$$M(\mathcal{R}h)(x) = M\left(\sum_{k=0}^{\infty} \frac{M^k h(x)}{2^k m^k} \right) \leqslant \sum_{k=0}^{\infty} \frac{M^{k+1} h(x)}{2^k m^k} = 2m \sum_{k=0}^{\infty} \frac{M^{k+1} h(x)}{2^{k+1} m^{k+1}}$$

$$\leqslant 2m \mathcal{R}h(x).$$

Let $Q \subset \mathbb{R}^n$ and $y \in Q$. Then

$$\fint_Q \mathcal{R}h(x)\,dx \leqslant M(\mathcal{R}h)(y) \leqslant 2m\mathcal{R}h(y).$$

Let us choose a sequence (y_j) such that $\mathcal{R}h(y_j) \to \operatorname{ess\,inf}_{x \in Q} \mathcal{R}h(x)$. Using this we obtain

$$\mathcal{R}h t_{A_1} = \sup_Q \left(\fint_Q \mathcal{R}h(x)\,dx \right) \operatorname{ess\,sup}_{x \in Q} \frac{1}{\mathcal{R}h(x)} \leqslant 2m \operatorname{ess\,sup}_{x \in Q} \frac{\mathcal{R}h(y_j)}{\mathcal{R}h(x)} \to 2m,$$

where we used that $h \neq 0$ implies $\mathcal{R}h(x) \geqslant \frac{1}{2m} Mh(x) \geqslant \frac{1}{2m} \fint_Q |h|\,dy > 0$ for every x. Thus we have $\mathcal{R}h \in A_1$ and $[\mathcal{R}h]_{A_1} \leqslant 2m$.

Fix $(f, g) \in \mathcal{F}$ and define $\mathcal{H} := \{h \in L^0(\mathbb{R}^n) : \|h\|_{L^{\psi_q^*}(\Omega)} \leqslant 1, \ h|_{\mathbb{R}^n \setminus \Omega} = 0\}$. By rescaling, the norm conjugate formula (Theorem 3.4.6) and (A),

$$\|f\|_{L^\psi(\Omega)}^q = \big\| |f|^q \big\|_{L^{\psi_q}(\Omega)} \approx \sup_{h \in \mathcal{H}} \int_\Omega |f|^q |h|\,dx \leqslant \sup_{h \in \mathcal{H}} \int_\Omega |f|^q \mathcal{R}h\,dx. \tag{5.3.2}$$

To apply our hypothesis, by our convention on families of extrapolation pairs we need to show that the left-hand side in (5.3.1) is finite. This follows from Hölder's inequality and (B): for all $h \in \mathcal{H}$,

$$\int_\Omega |f|^q \mathcal{R}h\,dx \leqslant 2\||f^q\|_{\psi_q} \|\mathcal{R}h\|_{\psi_q^*} \leqslant 4Q\|f\|_\psi^q \|h\|_{\psi_q^*} \leqslant 4Q\|f\|_\psi^q < \infty, \tag{5.3.3}$$

where the last inequality holds since, again by our convention, $f \in L^\psi$. Given this and (C) we can apply our hypothesis (5.3.1) in (5.3.2) to get that

$$\|f\|_{L^\psi(\Omega)} \leqslant \sup_{h \in \mathcal{H}} \left(\int_\Omega |f|^q \mathcal{R}h \, dx \right)^{\frac{1}{q}} \leqslant C \sup_{h \in \mathcal{H}} \left(\int_\Omega |g|^p (\mathcal{R}h)^{\frac{p}{q}} \, dx \right)^{\frac{1}{p}}.$$

Then for any $h \in \mathcal{H}$, by Hölder's inequality, rescaling (Lemma 5.2.1), and property (B),

$$\int_\Omega |g|^p (\mathcal{R}h)^{\frac{p}{q}}, dx \leqslant 2 \sup_{h \in \mathcal{H}} \left\| |g|^p \right\|_{\varphi_p} \left\| (\mathcal{R}h)^{\frac{p}{q}} \right\|_{\varphi_p^*}$$

$$\lesssim \sup_{h \in \mathcal{H}} \|g\|_\varphi^p \|\mathcal{R}h\|_{\psi_q^*}^{\frac{p}{q}} \lesssim \|g\|_\varphi^p. \tag{5.3.4}$$

We combine the last two inequalities and get $\|f\|_\psi \lesssim \|g\|_\varphi$, as desired. $\qquad\square$

We next apply the assumptions of Chap. 4 to get sufficient conditions on φ and ψ for the assumptions of Theorem 5.3.1 to hold. We also specialize from extrapolation pairs to operators. Note that Lemma 5.2.3 can be used to construct a suitable ψ from φ in the next result.

Theorem 5.3.2 (Off-Diagonal Extrapolation) *Let* $T : L^0(\Omega) \to L^0(\Omega)$ *and suppose that*

$$\|Tf\|_{L^q(\Omega,w)} \leqslant C([w]_{A_1})\|f\|_{L^p(\Omega,w^{p/q})}$$

for some p *and* q, *with* $1 \leqslant p \leqslant q < \infty$, *and all* $w \in A_1$. *Let* $\gamma := \frac{1}{p} - \frac{1}{q}$ *and* $\varphi, \psi \in \Phi_w(\Omega)$ *be such that* $\varphi^{-1}(x, t) \approx t^\gamma \psi^{-1}(x, t)$. *Let* $r \in (\gamma, \frac{1}{p}]$. *Assume that one of the following conditions holds:*

(a) φ *satisfies assumptions* (A0), (A1), (A2), (aInc)$_p$ *and* (aDec)$_{\frac{1}{r}}$,
(b) ψ *satisfies assumptions* (A0), (A1), (A2), (aInc)$_q$ *and* (aDec)$_{\frac{1}{r-\gamma}}$.

Then, for all $f \in L^\varphi(\Omega)$,

$$\|Tf\|_{L^\psi(\Omega)} \lesssim \|f\|_{L^\varphi(\Omega)}.$$

Proof We will derive this result as a consequence of the proof of Theorem 5.3.1. Define the family of extrapolation pairs

$$\mathcal{F} := \{(|Tg|, |g|) : g \in L^\varphi\}.$$

By inequality (5.3.4) we have that $g \in L^p(\Omega, (\mathcal{R}h)^{\frac{p}{q}})$ for every $h \in \mathcal{H}$. Since T maps $L^p(\Omega, w^{p/q})$ to $L^q(\Omega, w)$, we conclude that $Tg \in L^q(\Omega, \mathcal{R}h)$. This is our substitute for inequality (5.3.3) (with $f = |Tg|$) without the *a priori* assumption that $Tg \in L^\psi$. Therefore, the proof goes through and we get the desired conclusion once we have show that the maximal operator is bounded on $L^{\psi_q^*}$.

We note that the assumptions in (a) and (b) are equivalent: (A0), (A1) and (A2) are clear by the definitions since $\varphi^{-1}(x, t) \approx t^\gamma \psi^{-1}(x, t)$; (aInc)$_p$ and (aDec)$_{1/r}$ of φ imply (aInc)$_q$ and (aDec)$_{1/(r-\gamma)}$ of ψ as in Lemma 5.2.3; the other direction is similar. Therefore, it suffices to consider (b).

Since ψ satisfies (aInc)$_q$, we find that $\psi_q \in \Phi_w(\Omega)$. Since ψ satisfies (A0)–(A2) and (aDec)$_{1/(r-\gamma)}$, and since $\psi_q^* = T_q(\psi)^*$, ψ_q^* satisfies (A0)–(A2) and (aInc)$_{(1/q(r-\gamma))'}$ by Propositions 5.2.2 and 2.4.9, and Lemmas 3.7.6, 4.1.7 and 4.2.4. Since $\gamma < r \leqslant \frac{1}{p}$ we obtain that $(1/q(r - \gamma))' > 1$. Hence, by Theorem 4.3.4, the maximal operator is bounded on $L^{\psi_q^*}$.   □

Remark 5.3.3 In Theorem 5.3.2, we assume that T is defined on L^0 and that Tf is a measurable function. However, we do not assume that T is linear or sublinear.

In the special case $p = q$, we have the following corollary. We include also the statement of the vector-values inequality.

Corollary 5.3.4 *Given a family of extrapolation pairs $\mathcal{F}$, suppose that for some $p \in [1, \infty)$ and all $w \in A_p$,*

$$\|f\|_{L^p(\Omega, w)} \leqslant C([w]_{A_p})\|g\|_{L^p(\Omega, w)}, \qquad (f, g) \in \mathcal{F}. \tag{5.3.5}$$

Suppose $\varphi \in \Phi_w(\Omega)$ satisfies (A0), (A1), (A2) and (aDec). If $p > 1$, then we also assume (aInc). Then

$$\|f\|_{L^\varphi(\Omega)} \lesssim \|g\|_{L^\varphi(\Omega)}, \qquad (f, g) \in \mathcal{F}. \tag{5.3.6}$$

Moreover, for any $q \in (1, \infty)$,

$$\left\|\left(\sum_k |f_k|^q\right)^{1/q}\right\|_{L^\varphi(\Omega)} \lesssim \left\|\left(\sum_k |g_k|^q\right)^{1/q}\right\|_{L^\varphi(\Omega)}, \qquad \{(f_k, g_k)\}_k \subset \mathcal{F}. \tag{5.3.7}$$

Proof If $p > 1$, then φ satisfies (aInc)$_{p_-}$ for some $p_- > 1$ by assumption. By Theorem 5.1.1, (5.3.5) holds also for some $p \leqslant p_-$. If $p = 1$, then this is automatically true without using Theorem 5.1.1. Since $A_1 \subset A_p$, the hypotheses of Theorem 5.3.2 are satisfied, and so we get (5.3.6). To prove (5.3.7) we repeat this argument, starting from the weighted vector-valued inequality in Theorem 5.1.1.   □

Remark 5.3.5 An off-diagonal version of Corollary 5.3.4 holds, using the off-diagonal extrapolation theorem [30, Theorem 3.23]. Details are left to the interested reader.

It is possible to prove limited-range extrapolation results in the generalized Orlicz setting [29, Theorem 1.4], but we have not included them here.

5.4   Applications of Extrapolation

In this section, we give some representative applications of extrapolation to prove norm inequalities in generalized Orlicz spaces. The key to such inequalities is the existence of weighted norm inequalities and there is a vast literature on this subject. For additional examples in the context of variable exponent Lebesgue spaces that can be easily extended to generalized Orlicz spaces, see [27, 28, 30, 31].

To use extrapolation to prove norm inequalities we either need that the operator is *a priori* defined on L^φ or we need to use density and approximation arguments. In this case our conditions from the earlier chapters are very useful. For instance, if $\varphi \in \Phi_w(\Omega)$ satisfies (A0) and (aDec), then by Lemma 3.7.14 and Theorem 3.7.15 we obtain that $L_0^\infty(\Omega)$ and $C_0^\infty(\Omega)$ are both dense in $L^\varphi(\Omega)$.

To apply extrapolation via density and approximation, we consider a common special case. Suppose T is a linear operator that is defined on a dense subset $X \subset L^\varphi$, and suppose further that $X \subset L^p(\mathbb{R}^n, w)$ for all $p \geqslant 1$ and $w \in A_1$. (This is the case if $X = L_0^\infty$ or $X = C_0^\infty$.) If T satisfies weighted norm inequalities on $L^p(\mathbb{R}^n, w)$, then we can take as our extrapolation pairs the family

$$\mathcal{F} := \big\{(\min\{|Tf|, k\}\chi_{B(0,k)}, |f|) : f \in X\big\}.$$

By Corollary 3.7.10, L_0^∞ is contained in L^φ; hence, $\min\{|Tf|, k\}\chi_{B(0,k)} \in L^\varphi$ and so $\mathcal{F}$ satisfies the convention for families of extrapolation pairs. Thus, we can apply Theorem 5.3.1 (when $p = q$) to prove that for all $f \in X$, $\|Tf\|_\varphi \leqslant C\|f\|_\varphi$. Since T is linear, given an arbitrary $f \in L^\varphi$, if we take any sequence $(f_j) \subset X$ converging to f, (Tf_j) is Cauchy and we can define Tf as the limit. This extends the norm inequality to all of L^φ.

If T is not linear, then this argument does not work. However, suppose T has the property that $|Tf(x)| \leqslant T(|f|)(x)$ and if f is non-negative, and (f_j) is any non-negative sequence that increases point-wise to f, then

$$Tf(x) \leqslant \liminf_{k \to \infty} Tf_j(x).$$

(This is the case, for instance, if T is the maximal operator.) Then the above argument can essentially be repeated, since given non-negative $f \in L^\varphi$, $f_j = \min(f, k)\chi_{B(0,k)} \in L^\varphi \cap L_0^\infty$.

In the following examples we will state our hypotheses in terms of the assumptions used in Corollary 5.3.4. The necessary families of extrapolation pairs can be constructed using the above arguments. (In the case of variable exponent spaces, several examples are worked out in detail in [27, Chapter 5].) Obviously, the weaker assumptions of Theorem 5.3.1 can be used.

Though we assume the boundedness of the maximal operator in order to apply extrapolation, one important consequence is that we get vector-valued inequalities for the maximal operator. For $p, q \in (1, \infty)$, $w \in A_p$, and a sequence $(f_k)_k \subset L^0$, we have

$$\left\| \left(\sum_k (Mf_k)^q \right)^{1/q} \right\|_{L^p(\Omega, w)} \lesssim \left\| \left(\sum_k |f_k|^q \right)^{1/q} \right\|_{L^p(\Omega, w)},$$

see [6]. Therefore, we have the following result by Corollary 5.3.4.

Corollary 5.4.1 (Maximal Operator) *Suppose that* $\varphi \in \Phi_w(\Omega)$ *satisfies* (A0), (A1), (A2), (aInc) *and* (aDec). *Then for* $q \in (1, \infty)$,

$$\left\| \left(\sum_k (Mf_k)^q \right)^{1/q} \right\|_{L^\varphi(\Omega)} \lesssim \left\| \left(\sum_k |f_k|^q \right)^{1/q} \right\|_{L^\varphi(\Omega)}.$$

Note how the extrapolation method led to the extraneous assumption (aDec), which is not expected in such a result. We believe that this inequality can be proved without this assumption by using some other method.

Question 5.4.2 Does Corollary 5.4.1 hold without the assumption (aDec)?

Let us next study a bounded linear operator $T : L^2 \to L^2$ that satisfies the following conditions. Let Δ be the diagonal in $\mathbb{R}^n \times \mathbb{R}^n$, that is, $\Delta := \{(x, x) : x \in \mathbb{R}^n\}$. Assume that there exists a kernel $K : (\mathbb{R}^n \times \mathbb{R}^n) \setminus \Delta \to \mathbb{R}$ such that for all $f \in C_0^\infty$ and $x \notin \mathrm{spt}(f)$,

$$Tf(x) = \int_{\mathbb{R}^n} K(x, y) f(y) \, dy,$$

and, moreover, for some $\varepsilon > 0$ the kernel satisfies

$$|K(x, y)| \lesssim \frac{1}{|x - y|^n}, \quad \text{and}$$

$$|K(x, y) - K(x, y + h)| + |K(x, y) - K(x + h, y)| \lesssim \frac{|h|^\varepsilon}{|x - y|^{n+\varepsilon}},$$

for $2|h| \leqslant |x - y|$. Then T is called a *Calderón–Zygmund singular integral operator*. Calderón–Zygmund singular integral operators satisfy weighted norm inequalities: for $p \in (1, \infty)$ and $w \in A_p$,

$$\|Tf\|_{L^p(\mathbb{R}^n, w)} \lesssim \|f\|_{L^p(\mathbb{R}^n, w)},$$

see [37]. Therefore, we get the following result by Corollary 5.3.4:

Corollary 5.4.3 (Calderón–Zygmund Singular Integrals) *Suppose that* $\varphi \in \Phi_w(\mathbb{R}^n)$ *satisfies* (A0), (A1), (A2), (aInc) *and* (aDec) *and let* T *be a Calderón–Zygmund singular integral operator. Then*

$$\|Tf\|_{L^\varphi(\mathbb{R}^n)} \lesssim \|f\|_{L^\varphi(\mathbb{R}^n)}$$

and, for $q \in (1, \infty)$,

$$\left\|\left(\sum_k |Tf_k|^q\right)^{1/q}\right\|_{L^\varphi(\mathbb{R}^n)} \lesssim \left\|\left(\sum_k |f_k|^q\right)^{1/q}\right\|_{L^\varphi(\mathbb{R}^n)}.$$

We can also extend the Coifman–Fefferman inequality [23] relating singular integrals and the Hardy–Littlewood maximal function. Given $w \in A_\infty$ and $p \in (0, \infty)$,

$$\|Tf\|_{L^p(\mathbb{R}^n, w)} \lesssim \|Mf\|_{L^p(\mathbb{R}^n, w)}.$$

By extrapolation we can extend this to generalized Orlicz spaces. Note that (aInc) is not needed in this result, since the previous inequality holds also for $p = 1$.

Corollary 5.4.4 *Suppose that* $\varphi \in \Phi_w(\mathbb{R}^n)$ *satisfies* (A0), (A1), (A2), *and* (aDec) *and let* T *be a Calderón–Zygmund singular integral operator. Then*

$$\|Tf\|_{L^\varphi(\mathbb{R}^n)} \lesssim \|Mf\|_{L^\varphi(\mathbb{R}^n)}.$$

For $\alpha \in (0, n)$, we define the *Riesz potential* (also referred to as the fractional integral operator) by

$$I_\alpha f(x) := \int_\Omega \frac{|f(y)|}{|x - y|^{n-\alpha}} \, dy.$$

The associated *fractional maximal operator* is defined by

$$M_\alpha f(x) = \sup_{r>0} |B(x, r)|^{\frac{\alpha}{n}-1} \int_{B(x,r)\cap\Omega} |f(y)| \, dy.$$

These operators satisfy the following weighted norm inequalities: for $w \in A_1$ and p, q such that $p \in (1, \frac{n}{\alpha})$ and $\frac{1}{p} - \frac{1}{q} = \frac{\alpha}{n}$,

$$\|I_\alpha f\|_{L^q(\Omega, w)} \lesssim \|f\|_{L^p(\Omega, w^{p/q})} \quad \text{and} \quad \|M_\alpha f\|_{L^q(\Omega, w)} \lesssim \|f\|_{L^p(\Omega, w^{p/q})}.$$

(These inequalities are usually stated in terms of the A_{pq} condition of Muckenhoupt and Wheeden, but this special case is sufficient for our purposes. See [27] for further details.)

Therefore, by extrapolation (Theorem 5.3.2) we get the following results. The assumptions (a) and (b) in the next corollary are equivalent as noted in the proof of Theorem 5.3.2.

Corollary 5.4.5 (Riesz Potential and Fractional Maximal Operator) *Let* $\alpha \in (0, n)$ *and suppose* $\varphi, \psi \in \Phi_w(\Omega)$ *are such that* $\varphi^{-1}(x, t) \approx t^{\frac{\alpha}{n}} \psi^{-1}(x, t)$. *Let* $p, q \geqslant 1$ *with* $\frac{\alpha}{n} = \frac{1}{p} - \frac{1}{q}$ *and* $r \in (\frac{\alpha}{n}, \frac{1}{p}]$. *Assume that one of the following conditions holds:*

(a) φ *satisfies assumptions* (A0), (A1), (A2), (aInc)$_p$ *and* (aDec)$_{\frac{1}{r}}$.
(b) ψ *satisfies assumptions* (A0), (A1), (A2), (aInc)$_q$ *and* (aDec)$_{\frac{1}{r-\gamma}}$.

Then

$$\|I_\alpha f\|_{L^\psi(\Omega)} \lesssim \|f\|_{L^\varphi(\Omega)} \quad and \quad \|M_\alpha f\|_{L^\psi(\Omega)} \lesssim \|f\|_{L^\varphi(\Omega)}.$$

The fractional maximal operator can be estimated point-wise by the Riesz potential, $M_\alpha f \leqslant c(n) I_\alpha f$. There is no point-wise inequality in a opposite direction. However, for $w \in A_\infty$ and $p \in (0, \infty)$ we have the Coifman–Fefferman [23] inequality

$$\|I_\alpha f\|_{L^p(\mathbb{R}^n, w)} \lesssim \|M_\alpha f\|_{L^p(\mathbb{R}^n, w)}.$$

This inequality yields by Corollary 5.3.4 the following result without (aInc).

Corollary 5.4.6 *Suppose that* $\varphi \in \Phi_w(\mathbb{R}^n)$ *satisfies* (A0), (A1), (A2) *and* (aDec) *and* $\alpha \in (0, n)$. *Then* $\|I_\alpha f\|_{L^\varphi(\mathbb{R}^n)} \lesssim \|M_\alpha f\|_{L^\varphi(\mathbb{R}^n)}$.

5.5 Complex Interpolation

In this section we prove a complex interpolation theorem in the scale of generalized Orlicz spaces. Note that real interpolation has usually not been useful even in the variable exponent setting, since the primary and secondary parameter (i.e. p and θ in $(A, B)_{p,\theta}$) do not co-vary (but see [5] for an exception). Therefore, it is natural to start with complex interpolation also in more general generalized Orlicz spaces.

Previously, Musielak [96, Theorem 14.16] proved complex interpolation results, but his proofs were longer and more complicated; a simpler proof was given in [35]. However, in both cases the results only apply to N-functions which are proper. Here we eliminate these restrictions, proceeding along the lines of [34] and [29].

We recall the definition of the norm in the interpolation space $[L^{\varphi_0}, L^{\varphi_1}]_{[\theta]}$. Let $S := \{z \in \mathbb{C} : 0 < \mathrm{Re}\, z < 1\}$, so that $\overline{S} = \{z \in \mathbb{C} : 0 \leqslant \mathrm{Re}\, z \leqslant 1\}$, where $\mathrm{Re}\, z$ is the real part of z. Let $\mathcal{G}$ be the space of functions on $\overline{S}$ with values in $L^{\varphi_0} + L^{\varphi_1}$ which are holomorphic on S and bounded and continuous on $\overline{S}$ such that $F(it)$ and $F(1 + it)$ tend to zero as $|t| \to \infty$. (Recall that i denotes the imaginary unit. Also,

F is holomorphic with values in a Banach space means that $\frac{d}{d\bar{z}}F = 0$ in the Banach space.) For $F \in \mathcal{G}$ we set

$$\|F\|_{\mathcal{G}} := \sup_{t \in \mathbb{R}} \max \left\{ \|F(it)\|_{\varphi_0}, \|F(1+it)\|_{\varphi_1} \right\}.$$

Then we define the norm of $[L^{\varphi_0}, L^{\varphi_1}]_{[\theta]}$ for $f \in L^0$ by

$$\|f\|_{[\theta]} := \inf \left\{ \|F\|_{\mathcal{G}} : F \in \mathcal{G} \text{ and } f = F(\theta) \right\}.$$

For $\varphi_0, \varphi_1 \in \Phi_{\mathrm{w}}(\Omega)$ and $\theta \in (0, 1)$ we define the θ-intermediate Φ-function φ_θ by

$$\varphi_\theta^{-1} := (\varphi_0^{-1})^{1-\theta}(\varphi_1^{-1})^{\theta}.$$

Then φ_θ is also a weak Φ-function by Proposition 2.5.14. We use also this formula in the case $\theta \in \mathbb{C}$, however, then the left-hand side is not necessarily the left-inverse of any function.

We define the *right-continuous inverse* of $\varphi \in \Phi_{\mathrm{w}}$ by

$$\varphi^{\mathrm{r\text{-}inv}}(x, \tau) := \sup\{t \geqslant 0 : \varphi(x, t) \leqslant \tau\}.$$

As in Lemma 2.3.9, we can prove that $\varphi^{\mathrm{r\text{-}inv}}(\varphi(t)) \geqslant t$. Note that this inequality is the opposite from the case of the left-inverse.

Theorem 5.5.1 (Complex Interpolation) *Let $\varphi_0, \varphi_1 \in \Phi_{\mathrm{w}}(\Omega)$. Then*

$$\left[L^{\varphi_0}(\Omega), L^{\varphi_1}(\Omega) \right]_{[\theta]} = L^{\varphi_\theta}(\Omega) \quad \text{for all} \quad 0 < \theta < 1.$$

Proof By Theorem 2.5.10, we may assume without loss of generality that $\varphi_0, \varphi_1 \in \Phi_{\mathrm{s}}(\Omega)$. We extend all Φ-functions in this proof to complex numbers as follows: $\varphi : \Omega \times [0, \infty) \to [0, \infty]$ becomes $\varphi : \Omega \times \mathbb{C} \to [0, \infty]$ via $\varphi(x, t) = \varphi(x, |t|)$. For $z \in \mathbb{C}$ with $0 \leqslant \operatorname{Re} z \leqslant 1$ define ψ_z by

$$\psi_z(x, t) = \left(\varphi_0^{-1}(x, t) \right)^{1-z} \left(\varphi_1^{-1}(x, t) \right)^{z}.$$

Then $z \mapsto \psi_z$ is holomorphic on S and continuous on $\bar{S}$. When $z \in [0, 1]$, ψ_z is the left-inverse of some left-continuous weak Φ-function, $\psi_z = \varphi_z^{-1}$ (Proposition 2.5.14).

For $g \in L^{\varphi_\theta}(\Omega)$ with $\|g\|_{\varphi_\theta} \leqslant 1$, we define

$$f_\varepsilon(z, x) := \exp(-\varepsilon + \varepsilon z^2 - \varepsilon \theta^2) \, \psi_z \big(x, \varphi_\theta(x, g(x)) \big) \operatorname{sgn} g(x),$$

where $\operatorname{sgn} g$ is the sign of the function g. Note that $f_\varepsilon(it, x) \to 0$ and $f_\varepsilon(1 + it, x) \to 0$ as $|t| \to \infty$. Furthermore, $f_\varepsilon(\theta, x) = \exp(-\varepsilon)g(x)$ when $\varphi_\theta(x, g(x)) \in$

$(0, \infty)$ by Corollary 2.3.4 and

$$|f_\varepsilon(1 + it, x)| = \exp(-\varepsilon t^2 - \varepsilon \theta^2) \left| \psi_{1+it}(x, \varphi_\theta(x, g)) \right| \leqslant \varphi_1^{-1}(x, \varphi_\theta(x, g)),$$

$$|f_\varepsilon(it, x)| = \exp(-\varepsilon - \varepsilon t^2 - \varepsilon \theta^2) \left| \psi_{it}(x, \varphi_\theta(x, g)) \right| \leqslant \varphi_0^{-1}(x, \varphi_\theta(x, g)).$$

Since $\varphi_1(\varphi_1^{-1}(t)) \leqslant t$ (Lemma 2.3.9) and $\int_\Omega \varphi_\theta(x, g(x)) \, dx \leqslant 1$ (by the unit ball property) we conclude that $\varrho_{\varphi_1}(f_\varepsilon(1 + it, \cdot)) \leqslant 1$, similarly for φ_0. Thus

$$\|f_\varepsilon\|_{\mathcal{G}} = \sup_{t \in \mathbb{R}} \max \left\{ \|f_\varepsilon(it, \cdot)\|_{\varphi_0}, \|f_\varepsilon(1 + it, \cdot)\|_{\varphi_1} \right\} \leqslant 1.$$

This and $f_\varepsilon(\theta, x) = \exp(-\varepsilon) g(x)$ imply that $\|\exp(-\varepsilon) g\|_{[\theta]} \leqslant 1$. Letting $\varepsilon \to 0^+$, we find by a scaling argument that $\|g\|_{[\theta]} \leqslant \|g\|_{\varphi_\theta}$.

We now prove the opposite inequality, and start with several estimates. With the right-continuous inverses, we define

$$\xi_z(x, t) := c_\xi [(\varphi_0^*)^{\text{r-inv}}(x, t)]^{1-z} [(\varphi_1^*)^{\text{r-inv}}(x, t)]^z,$$

where the constant c_ξ will be specified shortly. By Lemma 2.5.8, $\varphi_0^* \in \Phi_c(\mathbb{R}^n)$ and hence it is strictly increasing when $\varphi_0^*(x, t) \in (0, \infty)$. Thus the right-continuous inverse agrees with the left-inverse, except possibly at the origin:

$$(\varphi_0^*)^{\text{r-inv}}(x, t) = \begin{cases} (\varphi_0^*)^{-1}(x, t) & \text{if } t > 0, \\ t_0(x) & \text{if } t = 0. \end{cases}$$

Here $t_0(x) := \sup\{t \geqslant 0 : \varphi_0^*(x, t) = 0\}$; similarly for φ_1^*. Then, for $\eta \in [0, 1]$ and $t > 0$, we obtain by Theorem 2.4.8 that

$$\xi_\eta(x, t) \approx \left[\frac{t}{(\varphi_0)^{-1}(x, t)} \right]^{1-\eta} \left[\frac{t}{(\varphi_1)^{-1}(x, t)} \right]^\eta$$

$$= \frac{t}{[(\varphi_0)^{-1}(x, t)]^{1-\eta} [(\varphi_1)^{-1}(x, t)]^\eta} = \frac{t}{\varphi_\eta^{-1}(x, t)} \qquad (5.5.1)$$

$$\approx (\varphi_\eta^*)^{-1}(x, t).$$

The constant $c_\xi \in (0, 1]$ is chosen to take care of the $\approx$-signs above, such that $\xi_\eta(x, t) \leqslant (\varphi_\eta^*)^{-1}(x, t)$ for $t > 0$. Furthermore, at the origin

$$\xi_\eta(x, 0) = c_\xi t_0(x)^{1-\eta} t_1(x)^\eta = c_\xi (\varphi_\eta^*)^{\text{r-inv}}(x, 0)$$

Assume now that $\theta \in (0, 1)$ and $\|g\|_{[\theta]} < 1$. By definition of $\|\cdot\|_{[\theta]}$, there exists $F : \overline{S} \to (L^{p_0(\cdot)} + L^{p(\cdot)})$ such that F is holomorphic on S and continuous on $\overline{S}$ with

$\|F\|_{\mathcal{F}} < 1$, $F(it)$ and $F(1 + it)$ tend to zero for $|t| \to \infty$, and $F(\theta) = g$. So the unit ball property implies $\sup_{t \in \mathbb{R}} \max \{ \varrho_{\varphi_0}(F(it)), \varrho_{\varphi_1}(F(1 + it)) \} \leqslant 1$.

Let $b \in L^{\varphi_\theta^*}$ with $\|b\|_{\varphi_\theta^*} \leqslant 1$ and define

$$h_\varepsilon(z, x) := \exp(-\varepsilon + \varepsilon z^2 - \varepsilon \theta^2) \, \xi_z\big(x, \varphi_\theta^*(x, b(x))\big) \, \mathrm{sgn} \, g(x).$$

Writing $F_\varepsilon(z) := \int_\Omega F(z, x) h_\varepsilon(z, x) \, dx$, we find by Young's inequality (2.4.1) that

$$|F_\varepsilon(it)| \leqslant \int_\Omega \xi_{it}\big(x, \varphi_\theta^*(x, b(x))\big) F(it, x) \, dx$$

$$\leqslant \int_\Omega \varphi_0^*\big[x, \xi_{it}\big(x, \varphi_\theta^*(x, b(x))\big)\big] + \varphi_0(x, F(it, x)) \, dx.$$

By the definition of ξ and (5.5.1),

$$|\xi_{it}(x, s)| = |\xi_0(x, s)| \leqslant (\varphi_0^*)^{-1}(x, s),$$

for $s > 0$. Thus by Lemma 2.3.9, $\varphi_0^*(x, \xi_{it}(x, s)) \leqslant s$. If $s = 0$, then $\xi_0(x, 0) = c_\xi t_0(x) \leqslant t_0(x)$. Hence $\varphi_0^*(x, \xi_0(x, 0)) \leqslant \varphi_0^*(x, t_0(x)) = 0$, by the left-continuity of φ_0^* and the definition of t_0. Thus in all cases $\varphi_0^*(x, \xi_{it}(x, s)) \leqslant s$, and the previous inequality gives $F_\varepsilon(it) \leqslant 2$. Analogously, $F_\varepsilon(1 + it) \leqslant 2$, so the three-line theorem [49] implies that $F_\varepsilon(z) \leqslant 2$ for all $z \in S$.

Next we consider the case $z = \theta$ so that $F(\theta, x) = g(x)$. Then $\xi_\theta(x, s) \approx (\varphi_\theta^*)^{-1}(x, s)$ for $s > 0$, by equation (5.5.1). Since $\|b\|_{\varphi_\theta^*} \leqslant 1$, we find that $\varphi_\theta^*(x, b(x)) = \infty$ in a set of measure zero. If $\varphi_\theta^*(x, b(x)) \in (0, \infty)$, then $(\varphi_\theta^*)^{-1}(x, \varphi_\theta^*(x, b(x))) = b(x)$ by Corollary 2.3.4. If $\varphi_\theta^*(x, b(x)) = 0$, then

$$\xi_\theta(x, \varphi_\theta^*(x, b(x))) = c_\xi (\varphi_\theta^*)^{\mathrm{r\text{-}inv}}(x, 0) \geqslant c_\xi b(x),$$

since $\varphi_\theta^*(x, s) = 0$ implies that $s \leqslant (\varphi_\theta^*)^{\mathrm{r\text{-}inv}}(x, 0)$. Thus, we obtain that

$$F_\varepsilon(\theta) = e^{-\varepsilon} \int_\Omega |g(x)| \, \xi_\theta\big(x, \varphi_\theta^*(x, b(x))\big) \, dx \geqslant c_\xi e^{-\varepsilon} \int_\Omega |g(x)| \, b(x) \, dx.$$

By Theorem 3.4.6 and this estimate, we obtain that

$$\|g\|_{\varphi_\theta} \lesssim \sup_{\|b\|_{\varphi_\theta^*} \leqslant 1} \int_\Omega |g| \, |b| \, dx \leqslant c_\xi^{-1} e^\varepsilon F_\varepsilon(\theta).$$

Thus $\|g\|_{\varphi_\theta} \leqslant c$ when $\|g\|_{[\theta]} < 1$; the general case $\|g\|_{\varphi_\theta} \lesssim \|g\|_{[\theta]}$ follows by a scaling argument. $\square$

Remark 5.5.2 Section 7.1 of [34] contains a proof of the complex interpolation theorem without the N-function assumption (for variable exponent Lebesgue spaces). However, that proof contains an error since it is based on the inequality $\varphi^{-1}(\varphi(t)) \geqslant t$, which is in general false. This problem is overcome above by the use of the right-continuous inversse.

The following result is proved using Theorem 5.5.1 by means of the Riesz–Thorin Interpolation Theorem and the Hahn–Banach Theorem, cf. [34, Corollary 7.1.4] and [35, Corollary A.5].

Corollary 5.5.3 *Let $\varphi_0, \varphi_1 \in \Phi_w(\mathbb{R}^n)$ and let T be a sublinear operator that is bounded from $L^{\varphi_j}(\Omega)$ to $L^{\varphi_j}(\Omega)$ for $j = 0, 1$. Then for $\theta \in (0, 1)$, T is also bounded from $L^{\varphi_\theta}(\Omega)$ to $L^{\varphi_\theta}(\Omega)$.*

The next result is proved using Calderón's interpolation theorem, cf. [34, Corollary 7.1.6].

Corollary 5.5.4 *Let $\varphi_0, \varphi_1 \in \Phi_w(\mathbb{R}^n)$, let X be a Banach space and let T be a linear operator that is bounded from X to $L^{\varphi_0}(\Omega)$ and compact from X to $L^{\varphi_1}(\Omega)$. Then for $\theta \in (0, 1)$, T is also compact from X to $L^{\varphi_\theta}(\Omega)$.*

Chapter 6
Sobolev Spaces

In this chapter we study Sobolev spaces with generalized Orlicz integrability. We point out the novelties in this new setting and assume that the readers are familiar with classical Sobolev spaces. We refer to the books [3, 46, 88, 131]. In Sect. 6.1, we establish the properties from Chap. 3 also for Sobolev spaces. Then we study Poincaré and Sobolev–Poincaré inequalities in both norm and modular form in Sects. 6.2 and 6.3. Finally, we consider the density of smooth functions in the Sobolev space (Sect. 6.4).

6.1 Basic Properties

In this section we define Sobolev spaces and prove functional analysis-type properties. Let $\Omega \subset \mathbb{R}^n$ be an open set. We start by recalling the definition of weak derivatives.

Definition 6.1.1 Assume that $u \in L^1_{\text{loc}}(\Omega)$. Let $\alpha := (\alpha_1, \ldots, \alpha_n) \in \mathbb{N}^n_0$ be a multi-index. If there exists $g \in L^1_{\text{loc}}(\Omega)$ such that

$$\int_\Omega u \frac{\partial^{\alpha_1 + \cdots + \alpha_n} h}{\partial^{\alpha_1} x_1 \cdots \partial^{\alpha_n} x_n} \, dx = (-1)^{\alpha_1 + \cdots + \alpha_n} \int_\Omega hg \, dx$$

for all $h \in C^\infty_0(\Omega)$, then g is called a *weak partial derivative* of u of order α. The function g is denoted by $\partial_\alpha u$ or by $\frac{\partial^{\alpha_1 + \cdots + \alpha_n} u}{\partial^{\alpha_1} x_1 \cdots \partial^{\alpha_n} x_n}$. Moreover, we write ∇u to denote the *weak gradient* $\left(\frac{\partial u}{\partial x_1}, \ldots, \frac{\partial u}{\partial x_n} \right)$ of u and we abbreviate $\partial_j u$ for $\frac{\partial u}{\partial x_j}$ with $j = 1, \ldots, n$. More generally we write $\nabla^k u$ to denote the tensor with entries $\partial_\alpha u$, $|\alpha| = k$.

The classical derivatives of a function are its weak derivatives. Also, by definition, $\nabla u = 0$ almost everywhere in an open set where $u = 0$.

© Springer Nature Switzerland AG 2019
P. Harjulehto, P. Hästö, *Orlicz Spaces and Generalized Orlicz Spaces*,
Lecture Notes in Mathematics 2236,
https://doi.org/10.1007/978-3-030-15100-3_6

Definition 6.1.2 Let $\varphi \in \Phi_w(\Omega)$. The function $u \in L^\varphi(\Omega) \cap L^1_{loc}(\Omega)$ belongs to the *Sobolev space* $W^{k,\varphi}(\Omega)$, $k \in \mathbb{N}$, if its weak partial derivatives $\partial_\alpha u$ exist and belong to $L^\varphi(\Omega)$ for all $|\alpha| \leqslant k$. We define a semimodular on $W^{k,\varphi}(\Omega)$ by

$$\varrho_{W^{k,\varphi}(\Omega)}(u) := \sum_{0 \leqslant |\alpha| \leqslant k} \varrho_{L^\varphi(\Omega)}(\partial_\alpha u);$$

it induces a (quasi-)norm by

$$\|u\|_{W^{k,\varphi}(\Omega)} := \inf \left\{ \lambda > 0 : \varrho_{W^{k,\varphi}(\Omega)}\left(\frac{u}{\lambda}\right) \leqslant 1 \right\}.$$

We define local Sobolev spaces as usual:

Definition 6.1.3 A function u belongs to the *local Sobolev space* $W^{k,\varphi}_{loc}(\Omega)$ if $u \in W^{k,\varphi}(U)$ for every open set $U \subset\subset \Omega$.

Sobolev functions, as Lebesgue functions, are defined only up to measure zero and thus we identify functions that are equal almost everywhere. If the set Ω is obvious from the context, we abbreviate $\|u\|_{W^{k,\varphi}(\Omega)}$ by $\|u\|_{k,\varphi}$ and $\varrho_{W^{k,\varphi}(\Omega)}$ by $\varrho_{k,\varphi}$.

In the next theorem we assume that $L^\varphi(\Omega) \subset L^1_{loc}(\Omega)$: this is a natural restriction, since the derivative is defined only for functions in $L^1_{loc}(\Omega)$. Recall that (A0) implies $L^\varphi(\Omega) \subset L^1_{loc}(\Omega)$ by Lemma 3.7.7. Note also that in (d) we could use Δ^w_2 and ∇^w_2 instead of (aInc) and (aDec).

Theorem 6.1.4 *Let $\varphi \in \Phi_w(\Omega)$ be such that $L^\varphi(\Omega) \subset L^1_{loc}(\Omega)$.*

(a) *Then $W^{k,\varphi}(\Omega)$ is a quasi-Banach space.*
(b) *If $\varphi \in \Phi_c(\Omega)$, then $W^{k,\varphi}(\Omega)$ is a Banach space.*
(c) *If φ satisfies (aDec), then $W^{k,\varphi}(\Omega)$ is separable.*
(d) *If φ satisfies (aInc) and (aDec), then $W^{k,\varphi}(\Omega)$ is uniformly convex and reflexive.*

Proof We prove only the case $k = 1$, the proof for the general case is similar. We first show that the Sobolev spaces is a quasi-Banach/Banach space according as $\varphi \in \Phi_w(\Omega)/\varphi \in \Phi_c(\Omega)$. Using similar arguments as in the proof of Lemma 3.2.2, we obtain that $\| \cdot \|_{1,\varphi}$ is a quasinorm if $\varphi \in \Phi_w(\Omega)$ and a norm if $\varphi \in \Phi_c(\Omega)$.

It remains to prove completeness. For that let (u_i) be a Cauchy sequence in $W^{1,\varphi}(\Omega)$. We have to show that there exists $u \in W^{1,\varphi}(\Omega)$ such that $u_i \to u$ in $W^{1,\varphi}(\Omega)$ as $i \to \infty$. Since the Lebesgue space $L^\varphi(\Omega)$ is a quasi-Banach space (Theorem 3.3.7), there exist $u, g_1, \ldots, g_n \in L^\varphi(\Omega)$ such that $u_i \to u$ and $\partial_j u_i \to g_j$ in $L^\varphi(\Omega)$ for every $j = 1, \ldots, n$. Let $h \in C^\infty_0(\Omega)$. Since u_i is in $W^{1,\varphi}(\Omega)$ we have

$$\int_\Omega u_i \, \partial_j h \, dx = - \int_\Omega h \, \partial_j u_i \, dx.$$

It follows by Theorem 3.4.6 that $L^{\varphi^*} = (L^\varphi)'$. This and the assumption $L^\varphi \subset L^1_{\text{loc}}$ yield that $L^\infty \subset L^{\varphi^*}_{\text{loc}}$, and so $h, \partial_j h \in L^{\varphi^*}(\Omega)$. It follows from Hölder's inequality (Lemma 3.2.11) that $\|u_i \partial_j h - u \partial_j h\|_1 \leqslant 2\|\partial_j h\|_{\varphi^*} \|u - u_i\|_\varphi$ and similarly for $h \partial_j u_i - h g_j$. Hence

$$\int_\Omega u_i \, \partial_j h \, dx \to \int_\Omega u \, \partial_j h \, dx \quad \text{and} \quad \int_\Omega h \, \partial_j u_i \, dx \to \int_\Omega h \, g_j \, dx$$

as $i \to \infty$ and so $(g_1, \dots, g_n)$ is the weak gradient of u. It follows that $u \in W^{1,\varphi}(\Omega)$. By $u_i \to u$ and $\partial_j u_i \to g_j$ in $L^\varphi(\Omega)$ we obtain for all $\lambda > 0$ by Lemma 3.3.1 that $\varrho_\varphi(\lambda(u - u_i)) \to 0$ and $\varrho_\varphi(\lambda(g_j - \partial_j u_i)) \to 0$, $j = 1, \dots, n$. Thus $\varrho_{1,\varphi}(\lambda(u - u_i)) \to 0$ for all $\lambda > 0$ and hence $u_i \to u$ in $W^{1,\varphi}(\Omega)$ by Lemma 3.3.1.

By Theorem 3.5.2, $L^\varphi(\Omega)$ is separable if φ satisfies (aDec) and by Theorem 3.6.7, $L^\varphi(\Omega)$ is uniformly convex and reflexive if φ satisfies Δ_2^w and ∇_2^w. These properties depend only on the space, so we may assume that φ is convex to apply Proposition 1.3.5 regarding Banach spaces. By the mapping $u \mapsto (u, \nabla u)$, the space $W^{1,\varphi}(\Omega)$ is a closed subspace of $L^\varphi(\Omega) \times (L^\varphi(\Omega))^n$. Thus $W^{1,\varphi}(\Omega)$ is separable if φ satisfies (aDec), and uniformly convex and reflexive if φ satisfies Δ_2^w and ∇_2^w, by Proposition 1.3.5. By Lemma 2.2.6 and Corollary 2.4.11, these hold under (aInc) and (aDec). $\square$

Lemma 6.1.5 *Let $\varphi \in \Phi_w(\Omega)$. Then*

$$\|u\|_{W^{k,\varphi}(\Omega)} \approx \sum_{m=0}^k \big\| |\nabla^m u| \big\|_{L^\varphi(\Omega)} \quad \text{and} \quad \varrho_{W^{k,\varphi}(\Omega)}(u) \simeq \sum_{m=0}^k \varrho_{L^\varphi(\Omega)}(|\nabla^m u|).$$

Both implicit constants depend only on n and k. If φ satisfies (aDec), *then $\simeq$ can be replaced by $\approx$.*

Proof We abbreviate $\big\| |\nabla^m u| \big\|_\varphi$ as $\|\nabla^m u\|_\varphi$, $m \in \mathbb{N}$. First we prove the claim for norms. Clearly $\sum_{m=0}^k \|\nabla^m u\|_\varphi \leqslant (k+1)\|u\|_{k,\varphi}$. Let $\lambda_m > \|\nabla^m u\|_\varphi$, $m = 0, \dots, k$ and $\ell := \sum_{0 \leqslant |\alpha| \leqslant k} 1$. Then, by (aInc)$_1$ and the unit ball property (Lemma 3.2.3),

$$\varrho_{k,\varphi}\left(\frac{u}{a\ell \sum_{m=0}^k \lambda_m}\right) = \sum_{0 \leqslant |\alpha| \leqslant k} \varrho_\varphi\left(\frac{\partial_\alpha u}{a\ell \sum_{m=0}^k \lambda_m}\right) \leqslant \frac{1}{\ell} \sum_{0 \leqslant |\alpha| \leqslant k} \varrho_\varphi\left(\frac{\partial_\alpha u}{\sum_{m=0}^k \lambda_m}\right)$$

$$\leqslant \frac{1}{\ell} \sum_{0 \leqslant |\alpha| \leqslant k} \varrho_\varphi\left(\frac{\partial_\alpha u}{\lambda_m}\right) \leqslant \frac{1}{\ell} \sum_{0 \leqslant |\alpha| \leqslant k} 1 = 1$$

and thus $\|u\|_{k,\varphi} \lesssim \sum_{m=0}^k \|\nabla^m u\|_\varphi$.

Then we prove the claim for modulars. Set $K := \sum_{|\alpha|=k} 1$. By (aInc)$_1$ we obtain

$$
\varrho_{k,\varphi}(u) = \sum_{0 \leqslant |\alpha| \leqslant k} \int_\Omega \varphi(x, \partial_\alpha u)\, dx = \sum_{m=0}^{k} \sum_{|\alpha|=m} \int_\Omega \varphi(x, \partial_\alpha u)\, dx
$$

$$
\leqslant K \sum_{m=0}^{k} \int_\Omega \varphi(x, |\nabla^m u|)\, dx \leqslant \sum_{m=0}^{k} \int_\Omega \varphi(x, aK|\nabla^m u|)\, dx.
$$

For the opposite direction we need quasi-convexity of φ (Corollary 2.2.2). Let $K_m := \sum_{|\alpha|=m} 1$ so that $1 \leqslant K_m \leqslant K$. Then

$$
\sum_{m=0}^{k} \varrho_\varphi\!\left(\frac{\beta}{K}|\nabla^m u|\right) \leqslant \sum_{m=0}^{k} \int_\Omega \varphi\!\left(x, \frac{\beta}{K_m}|\nabla^m u|\right) dx
$$

$$
\leqslant \sum_{m=0}^{k} \int_\Omega \varphi\!\left(x, \frac{\beta}{K_m} \sum_{|\alpha|=m} |\partial_\alpha u|\right) dx
$$

$$
\leqslant \sum_{m=0}^{k} \frac{1}{K_m} \sum_{|\alpha|=m} \int_\Omega \varphi(x, |\partial_\alpha u|)\, dx
$$

$$
\leqslant \sum_{0 \leqslant |\alpha| \leqslant k} \int_\Omega \varphi(x, \partial_\alpha u)\, dx = \varrho_{k,\varphi}(u).
$$

If φ satisfies (aDec), then $\simeq$ yields $\approx$ by Lemma 2.1.10. $\square$

Lemma 6.1.6 *Let* $\varphi \in \Phi_{\mathrm{w}}(\Omega)$ *satisfy* (A0) *and* (aInc)$_p$, $p \in [1,\infty)$. *Then* $W^{k,\varphi}(\Omega) \hookrightarrow W^{k,p}_{\mathrm{loc}}(\Omega)$. *If* $|\Omega| < \infty$, *then* $W^{k,\varphi}(\Omega) \hookrightarrow W^{k,p}(\Omega)$.

Proof By Lemma 3.7.7, $L^\varphi(\Omega) \hookrightarrow L^p(\Omega) + L^\infty(\Omega) \hookrightarrow L^p(\Omega) + L^p_{\mathrm{loc}}(\Omega)$, where loc is not needed if $|\Omega| < \infty$. This yields the claim by Lemma 6.1.5. $\square$

A (real-valued) function space is a *lattice* if the point-wise minimum and maximum of any two of its elements belong to the space. The next lemma and Lemma 6.1.6 imply that $W^{1,\varphi}(\Omega)$ is a lattice provided $L^\varphi(\Omega) \subset L^1_{\mathrm{loc}}(\Omega)$

Lemma 6.1.7 (Theorem 1.20, [63]) *If* $u, v \in W^{1,1}_{\mathrm{loc}}(\Omega)$, *then* $\max\{u, v\}$ *and* $\min\{u, v\}$ *are in* $W^{1,1}_{\mathrm{loc}}(\Omega)$ *with*

$$
\nabla \max(u, v)(x) = \begin{cases} \nabla u(x), & \text{for almost every } x \in \{u \geqslant v\}; \\ \nabla v(x), & \text{for almost every } x \in \{v \geqslant u\}; \end{cases}
$$

and

$$\nabla \min(u, v)(x) = \begin{cases} \nabla u(x), & \text{for almost every } x \in \{u \leqslant v\}; \\ \nabla v(x), & \text{for almost every } x \in \{v \leqslant u\}. \end{cases}$$

In particular, $|u|$ belongs to $W^{1,1}_{\text{loc}}(\Omega)$ and $|\nabla|u|| = |\nabla u|$ almost everywhere in Ω.

We close this section by defining Sobolev spaces with zero boundary values and establishing basic properties for them.

Definition 6.1.8 Let $\varphi \in \Phi_w(\Omega)$ and $k \in \mathbb{N}$. The *Sobolev space $H^{k,\varphi}_0(\Omega)$ with zero boundary values* is the closure of $C^\infty_0(\Omega) \cap W^{k,\varphi}(\Omega)$ in $W^{k,\varphi}(\Omega)$.

Since $H^{k,\varphi}_0(\Omega)$ is a closed subspace of $W^{k,\varphi}(\Omega)$, the following theorem follows by Proposition 1.3.5 and Theorem 6.1.4.

Theorem 6.1.9 *Let $\varphi \in \Phi_w(\Omega)$ be such that $L^\varphi(\Omega) \subset L^1_{\text{loc}}(\Omega)$. Then $H^{k,\varphi}_0(\Omega)$ is a quasi-Banach space (Banach, if $\varphi \in \Phi_c(\Omega)$). $H^{k,\varphi}_0(\Omega)$ is separable if φ satisfies* (aDec), *and uniformly convex and reflexive if φ satisfies* (aInc) *and* (aDec).

Note that our definition for zero boundary Sobolev spaces is not that reasonable if smooth functions are not dense in $W^{k,\varphi}(\Omega)$. Let $u \in W^{k,\varphi}(\Omega)$ and let $v \in C_0(\Omega)$ be Lipschitz-continuous. Then $uv \in W^{k,\varphi}(\Omega)$ has a compact support in Ω but it might happen that $uv \notin H^{1,\varphi}_0(\Omega)$. In the case $\varphi(x, t) = t^{p(x)}$, this question have been studied in [34, Section 11.2].

Lemma 6.1.10 *Let $\varphi \in \Phi_w(\Omega)$ be such that $C^\infty(\Omega) \cap W^{1,\varphi}(\Omega)$ is dense in $W^{1,\varphi}(\Omega)$. If $u \in W^{1,\varphi}(\Omega)$ and $\operatorname{spt} u \subset\subset \Omega$, then $u \in H^{1,\varphi}_0(\Omega)$.*

Proof Let (ξ_i) be a sequence in $C^\infty(\Omega) \cap W^{1,\varphi}(\Omega)$ converging to u in $W^{1,\varphi}(\Omega)$. Let $\eta \in C^\infty_0(\Omega)$ be a cut-off function with $0 \leqslant \eta \leqslant 1$ and $\eta \equiv 1$ in $\operatorname{spt} u$. Then $(\eta \xi_i)$ is a sequence in $C^\infty_0(\Omega)$. Since $u - \eta \xi_i = \eta(u - \xi_i)$ we obtain

$$\|u - \eta \xi_i\|_\varphi \leqslant \|\eta\|_{L^\infty} \|u - \xi_i\|_\varphi$$

and

$$\|\nabla(u - \eta \xi_i)\|_\varphi \leqslant \|\nabla \eta\|_{L^\infty} \|u - \xi_i\|_\varphi + \|\eta\|_{L^\infty} \|\nabla u - \nabla \xi_i\|_\varphi$$

and thus by Lemma 6.1.5 we have $\eta \xi_i \to u$ in $W^{1,\varphi}(\Omega)$, and furthermore $u \in H^{1,\varphi}_0(\Omega)$. $\square$

Note that, by Theorem 6.4.7, $C^\infty(\Omega) \cap W^{1,\varphi}(\Omega)$ is dense in $W^{1,\varphi}(\Omega)$ if $\varphi \in \Phi_w(\Omega)$ satisfies (A0), (A1), (A2) and (aDec).

6.2 Poincaré Inequalities

Poincaré inequalities hold in sufficiently regular domains. Concerning the regularity of the domain we focus here on bounded John domains. The L^p-Poincaré inequality is known to hold for more irregular domains but the inequality is mostly used in John domains.

Definition 6.2.1 A bounded domain $\Omega \subset \mathbb{R}^n$ is called an α-*John domain*, $\alpha > 0$, if there exists $x_0 \in \Omega$ (the *John center*) such that each point in Ω can be joined to x_0 by a rectifiable path γ (the *John path*) parametrized by its arc-length such that

$$B\big(\gamma(t), \tfrac{1}{\alpha}t\big) \subset \Omega$$

for all $t \in [0, \ell(\gamma)]$, where $\ell(\gamma) \leqslant \alpha$ is the length of γ. The ball $B(x_0, \frac{\mathrm{diam}(\Omega)}{2\alpha})$ is called the *John ball*.

Example 6.2.2 All bounded convex domains and bounded domains with Lipschitz boundary are John domains. But also $B(0, 1) \setminus [0, e_1) \subset \mathbb{R}^2$ is a John domain. John domains may possess fractal boundaries or internal cusps with a zero angle while external cusps with a zero angle are excluded. For example, the interior of Koch's snowflake is a John domain.

We recall the following well-known lemma that estimates u in terms of the Riesz potential, see [118] or [12, Chapter 6]. For this formulation a proof can be found in [34, Lemma 8.2.1]. Recall that I_1 denotes the Riesz potential operator (cf. Sect. 5.4).

Lemma 6.2.3

(a) *For every* $u \in H_0^{1,1}(\Omega)$,

$$|u| \lesssim I_1 |\nabla u| \quad \text{a.e. in } \Omega,$$

where the constant depends only on the dimension n.
(b) *If* Ω *is a John domain, then there exists a ball* $B \subset \Omega$ *such that*

$$|u(x) - u_B| \lesssim I_1 |\nabla u|(x) \quad \text{a.e. } x \in \Omega$$

for every $u \in W^{1,1}(\Omega)$. *The ball* B *satisfies* $|B| \approx |\Omega|$ *and the constants depend only on the dimension* n *and John-constant* α.

Remark 6.2.4 Similarly one may prove that the assertions in Lemma 6.2.3(b) also hold for $u \in L_{\mathrm{loc}}^1(\Omega)$ with $|\nabla u| \in L^1(\Omega)$.

We need the following lemma which allows us to change the set in the integral average.

Lemma 6.2.5 *Let Ω be a John domain and let $\varphi \in \Phi_{\mathrm{w}}(\Omega)$ satisfy (A0) and (A1). If $A \subset \Omega$ has positive finite measure, then*

$$\frac{|A|}{|\Omega|}\|u - u_A\|_{L^\varphi(\Omega)} \lesssim \|u - u_\Omega\|_{L^\varphi(\Omega)} \lesssim \|u - u_A\|_{L^\varphi(\Omega)}$$

for $u \in L^\varphi(\Omega)$. Here the constants depend on the John constant α.

Proof Let us first prove the upper bound. By the triangle inequality,

$$\|u - u_\Omega\|_{L^\varphi(\Omega)} \lesssim \|u - u_A\|_{L^\varphi(\Omega)} + \|u_A - u_\Omega\|_{L^\varphi(\Omega)}.$$

We estimate the second term by Hölder's inequality:

$$\|u_A - u_\Omega\|_{L^\varphi(\Omega)} = |u_A - u_\Omega|\,\|1\|_{L^\varphi(\Omega)} = |\Omega|^{-1}\|u - u_A\|_{L^1(\Omega)}\|1\|_{L^\varphi(\Omega)}$$

$$\leqslant c\,\frac{\|1\|_{L^{\varphi^*}(\Omega)}\|1\|_{L^\varphi(\Omega)}}{|\Omega|}\|u - u_A\|_{L^\varphi(\Omega)}.$$

Since Ω is bounded, (A2) follows by Lemma 4.2.3. Since Ω is a John domain, there exist balls $B \subset \Omega \subset B'$ with $|B| \approx |B'|$, where the implicit constant depends on the John constant. Lemma 4.4.8 yields that $\|1\|_{L^{\varphi^*}(\Omega)}\|1\|_{L^\varphi(\Omega)} \approx |\Omega|$ and hence the upper bound follows.

Then we move to the lower bound. By the triangle inequality,

$$\|u - u_A\|_{L^\varphi(\Omega)} \leqslant \|u - u_\Omega\|_{L^\varphi(\Omega)} + \|u_A - u_\Omega\|_{L^\varphi(\Omega)}.$$

We estimate the second term:

$$\|u_A - u_\Omega\|_{L^\varphi(\Omega)} = |u_A - u_\Omega|\,\|1\|_{L^\varphi(\Omega)} = |A|^{-1}\|u - u_\Omega\|_{L^1(\Omega)}\|1\|_{L^\varphi(\Omega)}.$$

The rest of the proof is like in the previous case.   □

Our arguments are based on the following averaging operator calculated over dyadic cubes.

Definition 6.2.6 For $k \in \mathbb{Z}$ we define the *averaging operator T_k* over dyadic cubes by

$$T_k f(x) := \sum_{\substack{Q \text{ dyadic} \\ \mathrm{diam}(Q)=2^{-k}}} \frac{\chi_{Q\cap\Omega}(x)}{|Q|} \int_{2Q\cap\Omega} f(y)\,dy$$

for all $f \in L^1_{\mathrm{loc}}(\mathbb{R}^n)$.

The following result is [34, Lemma 6.1] but we include the short proof for the reader's convenience.

Lemma 6.2.7 *Let $x \in \Omega$, $\delta > 0$, $\alpha \in (0, n)$, and $f \in L^1_{loc}(\Omega)$. Then*

$$\int_{B(x,\delta)\cap\Omega} \frac{|f(y)|}{|x-y|^{n-\alpha}} \, dy \lesssim \delta^\alpha \sum_{k=0}^{\infty} 2^{-\alpha k} T_{k+k_0} f(x) \lesssim \delta^\alpha M f(x),$$

where $k_0 \in \mathbb{Z}$ is chosen such that $2^{-k_0-1} \leqslant \delta < 2^{-k_0}$.

Proof We split the integration domain into annuli and use the definition of T_k:

$$\int_{B(x,\delta)\cap\Omega} \frac{|f(y)|}{|x-y|^{n-\alpha}} \, dy \leqslant \sum_{k=1}^{\infty} \left(\delta 2^{-k}\right)^{\alpha-n} \int_{\{2^{-k}\delta\leqslant|x-y|<2^{-k+1}\delta\}\cap\Omega} |f(y)| \, dy$$

$$\lesssim \delta^\alpha \sum_{k=0}^{\infty} 2^{-\alpha k} T_{k+k_0} f(x),$$

where we used the fact that $B(x, 2^{-k+1}\delta)$ is covered by at most 4^n dyadic cubes of size 2^{-k-k_0}. This is the first inequality. Since $T_{k+k_0} f \lesssim M f$, the second inequality follows by convergence of the geometric series. □

Note that the next result is established without the (aInc) and (aDec) assumptions.

Theorem 6.2.8 (Poincaré Inequality) *Let Ω be a bounded domain and let $\varphi \in \Phi_w(\Omega)$ satisfy (A0) and (A1).*

(a) *For every $u \in H_0^{1,\varphi}(\Omega)$, we have*

$$\|u\|_{L^\varphi(\Omega)} \lesssim \text{diam}(\Omega)\|\nabla u\|_{L^\varphi(\Omega)}.$$

(b) *If Ω is a John domain, then for every $u \in W^{1,\varphi}(\Omega)$, we have*

$$\|u - u_\Omega\|_{L^\varphi(\Omega)} \lesssim \text{diam}(\Omega)\|\nabla u\|_{L^\varphi(\Omega)}.$$

Proof We prove only the latter case. The proof for the first case is similar; the only difference is to use in Lemma 6.2.3 case (a) instead of case (b).

By Lemma 3.7.9, $W^{1,\varphi} \subset L^1_{loc}$. Thus, by Lemma 6.2.3(b) and Lemma 6.2.7, we have

$$|u(x) - u_B| \leqslant I_1 |\nabla u|(x) \lesssim \text{diam}(\Omega) \sum_{k=0}^{\infty} 2^{-k} T_{k+k_0} |\nabla u|(x)$$

for every $u \in W^{1,\varphi}(\Omega)$ and almost every $x \in \Omega$, where $k_0 \in \mathbb{Z}$ is chosen such that $2^{-k_0-1} \leqslant \text{diam}(\Omega) < 2^{-k_0}$. Since Ω is bounded, (A2) holds by Lemma 4.2.3. Thus we obtain by Theorem 4.4.3 that T_{k+k_0} is bounded. Using also the quasi-triangle inequality (Corollary 3.2.5) and convergence of a geometric series, we obtain

$$\|u - u_B\|_{L^\varphi(\Omega)} \lesssim \text{diam}(\Omega) \sum_{k=0}^{\infty} 2^{-k} \|T_{k+k_0}|\nabla u|\|_{L^\varphi(\Omega)}$$

$$\lesssim \text{diam}(\Omega)\|\nabla u\|_{L^\varphi(\Omega)}.$$

The estimate for $\|u - u_\Omega\|_{L^\varphi(\Omega)}$ follows from this and Lemma 6.2.5. $\square$

Theorem 6.2.8(a) immediately yields that $\|\nabla u\|_\varphi$ is an equivalent norm in $H_0^{1,\varphi}(\Omega)$:

Corollary 6.2.9 *Let Ω be a bounded domain and let $\varphi \in \Phi_w(\Omega)$ satisfy* (A0) *and* (A1). *For every $u \in H_0^{1,p(\cdot)}(\Omega)$,*

$$\|\nabla u\|_{L^\varphi(\Omega)} \leqslant \|u\|_{W^{1,\varphi}(\Omega)} \leqslant \left(1 + c \, \text{diam}(\Omega)\right) \|\nabla u\|_{L^\varphi(\Omega)}.$$

Let us next consider modular versions of the Poincaré inequality which are very important for applications to differential equations and calculus of variations. In the constant exponent case there is an obvious connection between modular and norm versions of the inequality, which does not hold even in the variable exponent context (see [34, Example 8.2.7]) and thus an extra "error term" is needed on the right hand side.

Proposition 6.2.10 (Poincaré Inequality in Modular Form) *Let Ω be a bounded domain and let $\varphi \in \Phi_w(\Omega)$ satisfy* (A0) *and* (A1).

(a) *There exists $\beta > 0$ independent of Ω such that*

$$\int_\Omega \varphi\left(x, \frac{\beta|u|}{\text{diam}(\Omega)}\right) dx \leqslant \int_\Omega \varphi(x, |\nabla u|) \, dx + \left|\{|\nabla u| \neq 0\} \cap \Omega\right|,$$

for every $u \in H_0^{1,1}(\Omega)$ with $\varrho_\varphi(|\nabla u|) \leqslant 1$.

(b) *If Ω is a John domain, then there exists $\beta > 0$ depending on the John constant such that*

$$\int_\Omega \varphi\left(x, \frac{\beta|u - u_\Omega|}{\text{diam}(\Omega)}\right) dx \leqslant \int_\Omega \varphi(x, |\nabla u|) \, dx + \left|\{|\nabla u| \neq 0\} \cap \Omega\right|,$$

for every $u \in W^{1,1}(\Omega)$ with $\varrho_\varphi(|\nabla u|) \leqslant 1$.

Proof Let us first note that since Ω is bounded, (A2) holds by Lemma 4.2.3. Hence (A0), (A1) and (A2) all hold.

We first prove (a). By Lemmas 6.2.3(a) and 6.2.7, we obtain

$$|u(x)| \leqslant c \int_\Omega \frac{|\nabla u(y)|}{|x-y|^{n-1}} dy \leqslant c_1 \operatorname{diam}(\Omega) \sum_{k=0}^\infty 2^{-k} T_{k+k_0} |\nabla u|(x),$$

where $k_0 \in \mathbb{Z}$ is chosen such that $2^{-k_0-1} \leqslant \operatorname{diam}(\Omega) < 2^{-k_0}$.

Let $\beta \in (0, 1]$ be from quasi-convexity of φ (Corollary 2.2.2), and β' and h from the key estimate (Theorem 4.3.2). We divide by $c_1 \operatorname{diam}(\Omega)$ and obtain by quasi-convexity of φ (Corollary 2.2.2) that

$$\int_\Omega \varphi\left(x, \frac{c_2 \beta \beta' |u|}{2c_1 \operatorname{diam}(\Omega)}\right) dx \leqslant \int_\Omega \varphi\left(x, c_2 \beta \beta' \sum_{k=0}^\infty 2^{-k-1} T_{k+k_0} |\nabla u|\right) dx$$

$$\leqslant \sum_{k=0}^\infty 2^{-k-1} \sum_{\substack{Q \text{ dyadic} \\ \operatorname{diam}(Q)=2^{-k}}} \int_{Q \cap \Omega} \varphi(x, c_2 \beta' T_{k+k_0} |\nabla u|) \, dx.$$

We fix c_2 in the next estimate. Let $R := 2^{-k-k_0}$ and Q_x be the dyadic cube of diameter R to which x belongs. Since $\varrho_\varphi(|\nabla u|) \leqslant 1$, the key estimate (Theorem 4.3.2), with $p = 1$, yields

$$\varphi\left(x, c_2 \beta' T_{k+k_0} |\nabla u|\right) \leqslant \varphi\left(x, \frac{\beta'}{|B(x, R)|} \int_{B(x,R) \cap \Omega} |\nabla u| \chi_{2Q_x} \, dy\right)$$

$$\leqslant \frac{1}{|B(x, R)|} \int_{B(x,R) \cap \Omega} \varphi(y, |\nabla u| \chi_{2Q_x}) + \|h\|_\infty \frac{|\{|\nabla u|>0\} \cap B(x,R)|}{|B(x,R)|}$$

$$\lesssim \frac{1}{|2Q_x|} \int_{2Q_x \cap \Omega} \varphi(y, |\nabla u|) \, dy + \frac{|\{|\nabla u|>0\} \cap 2Q_x|}{|2Q_x|}.$$

Combining this with the previous estimate, we obtain

$$\int_\Omega \varphi\left(x, \frac{c_2 \beta \beta' |u|}{2c_1 \operatorname{diam}(\Omega)}\right) dx$$

$$\lesssim \sum_{k=0}^\infty 2^{-k} \sum_{\substack{Q \text{ dyadic} \\ \operatorname{diam}(Q)=2^{-k}}} \int_{Q \cap \Omega} \frac{1}{|2Q|} \int_{2Q \cap \Omega} \varphi(y, |\nabla u|) \, dy + \frac{|\{|\nabla u|>0\} \cap 2Q|}{|2Q|} \, dx$$

$$\leqslant \sum_{k=0}^\infty 2^{-k} \sum_{\substack{Q \text{ dyadic} \\ \operatorname{diam}(Q)=2^{-k}}} \left(\int_{2Q \cap \Omega} \varphi(y, |\nabla u|) \, dy + |\{|\nabla u| > 0\} \cap 2Q|\right)$$

$$\lesssim \int_\Omega \varphi(y, |\nabla u|) \, dy + |\{|\nabla u| > 0\}|,$$

where we used in the last inequality that the dyadic cubes $2Q$ are locally N-finite. By (aInc)$_1$ constants from the right hand side can be absorbed to the left hand side.

Let us then move to (b). Instead of Lemma 6.2.3(a) we use Lemma 6.2.3(b) and obtain

$$|u(x) - u_B| \leqslant c \int_\Omega \frac{|\nabla u(y)|}{|x-y|^{n-1}} dy \leqslant c \operatorname{diam}(\Omega) \sum_{k=0}^\infty 2^{-k} T_{k+k_0} |\nabla u|(x), \quad (6.2.1)$$

where $k_0 \in \mathbb{Z}$ is chosen such that $2^{-k_0-1} \leqslant \operatorname{diam}(\Omega) < 2^{-k_0}$. Exactly as in (a), this yields that

$$\int_\Omega \varphi\left(x, \frac{\beta|u-u_B|}{\operatorname{diam}(\Omega)}\right) dx \leqslant \int_\Omega \varphi(x, |\nabla u|)\, dx + |\{|\nabla u| > 0\}|,$$

for every $u \in W^{1,1}(\Omega)$ with $\varrho_\varphi(|\nabla u|) \leqslant 1$.

Since $|u - u_\Omega| \leqslant |u - u_B| + |u_\Omega - u_B|$, we obtain

$$\int_\Omega \varphi\left(x, \frac{\frac{c_3\beta}{2}|u-u_\Omega|}{\operatorname{diam}(\Omega)}\right) dx \leqslant \int_\Omega \varphi\left(x, \frac{\beta|u-u_B|}{\operatorname{diam}(\Omega)}\right) dx$$
$$+ \int_\Omega \varphi\left(x, \frac{c_3\beta \fint_\Omega |u(y) - u_B|\, dy}{\operatorname{diam}(\Omega)}\right) dx.$$

So we need to estimate only the second term. By the $W^{1,1}$-Poincaré inequality,

$$\fint_\Omega |u(y) - u_B|\, dy \lesssim \operatorname{diam}(\Omega) \fint_\Omega |\nabla u|\, dy,$$

where $u \in W^{1,1}(\Omega)$. Note that this follows for example applying (6.2.1) in $W^{1,1}(\Omega)$. Let $\tilde{B}$ be a ball with radius $\operatorname{diam}(\Omega)$ containing Ω. Since Ω is a John domain we have $|\tilde{B}| \approx |\Omega|$. We fix c_3 in the next estimate. The $W^{1,1}$-Poincaré inequality and the key estimate (Theorem 4.3.2), with $p = 1$, yield

$$\int_\Omega \varphi\left(x, \frac{c_3\beta \fint_\Omega |u(y) - u_B|\, dy}{\operatorname{diam}(\Omega)}\right) dx \leqslant \int_\Omega \varphi\left(x, \beta \fint_\Omega |\nabla u|\, dy\right) dx$$
$$\leqslant \int_\Omega \varphi\left(x, \beta \fint_{\tilde{B}} \chi_\Omega |\nabla u|\, dy\right) dx$$
$$\leqslant \int_\Omega \frac{1}{|\tilde{B}|} \int_\Omega \varphi(y, |\nabla u|)\, dy + \frac{|\{|\nabla u|\neq 0\}|}{|\tilde{B}|}\, dx$$
$$\leqslant \int_\Omega \varphi(y, |\nabla u|)\, dy + |\{|\nabla u| \neq 0\}|.$$

Combining these estimates, we obtain (b). □

6.3 Sobolev Embeddings

Next we prove Sobolev–Poincaré inequalities in John domains. Note that for a given φ, a suitable function $\psi \in \Phi_w(\Omega)$ in the next theorem exist by Lemma 5.2.3.

Question 6.3.1 Does the next result hold without the assumption (aInc)?

Theorem 6.3.2 (Sobolev-Poincaré Inequality) *Let Ω be a John domain. Let $\varphi \in \Phi_w(\Omega)$ satisfy (A0), (A1), (A2), (aInc) and (aDec)$_q$, $q < n$. Suppose that $\psi \in \Phi_w(\Omega)$ satisfies $t^{-\frac{1}{n}} \varphi^{-1}(x, t) \approx \psi^{-1}(x, t)$. Then for every $u \in W^{1,\varphi}(\Omega)$,*

$$\|u - u_\Omega\|_{L^\psi(\Omega)} \lesssim \|\nabla u\|_{L^\varphi(\Omega)}.$$

Proof By Lemma 6.2.3(b),

$$|u(x) - u_B| \lesssim I_1|\nabla u|(x)$$

for every $u \in W^{1,\varphi}(\Omega)$ and almost every $x \in \Omega$. Thus

$$\|u - u_B\|_{L^\psi(\Omega)} \lesssim \|I_1|\nabla u|\|_{L^\psi(\Omega)}.$$

Corollary 5.4.5 yields $\|I_1|\nabla u|\|_{L^\psi(\Omega)} \lesssim \|\nabla u\|_{L^\varphi(\Omega)}$. Thus the estimate for $\|u - u_\Omega\|_{L^\psi(\Omega)}$ follows from Lemma 6.2.5.   □

We obtain the Sobolev embedding in John domains as a corollary.

Corollary 6.3.3 (Sobolev Embedding) *Let Ω be a John domain. Assume that $\varphi \in \Phi_w(\Omega)$ satisfies (A0), (A1), (A2), (aInc) and (aDec)$_q$, $q < n$. Suppose that $\psi \in \Phi_w(\Omega)$ satisfies $t^{-\frac{1}{n}} \varphi^{-1}(x, t) \approx \psi^{-1}(x, t)$. Then $W^{1,\varphi}(\Omega) \hookrightarrow L^\psi(\Omega)$.*

Proof By Theorem 6.3.2 and Hölder's inequality (Lemma 3.2.11) we obtain

$$\|u\|_\psi \lesssim \|u - u_\Omega\|_\psi + \|u_\Omega\|_\psi$$

$$\leqslant \|\nabla u\|_\varphi + |\Omega|^{-1}\|1\|_\psi\|u\|_1$$

$$\leqslant \|\nabla u\|_\varphi + 2|\Omega|^{-1}\|1\|_\psi\|1\|_{\varphi^*}\|u\|_\varphi.$$

The norms $\|1\|_\psi$ and $\|1\|_{\varphi^*}$ are finite by Corollary 3.7.10 and hence the claim follows.   □

We can use extrapolation to prove the Sobolev embedding theorem and then combine this with interpolation to prove a version of the Rellich–Kondratchov (compact embedding) theorem. We begin with the following weighted norm inequality: for all $f \in C_0^\infty(\Omega)$, $w \in A_1$, and $p \in [1, n)$,

$$\|f\|_{L^{p^*}(\Omega, w)} \leqslant C\|\nabla f\|_{L^p(\Omega, w^{p/p^*})},$$

where $p^* = \frac{np}{n-p}$ is the Sobolev exponent of p ([30, Lemma 4.31] or [27, Lemma 6.32]). We use extrapolation, Theorem 5.3.2, to obtain the following result with weaker assumptions than in Corollary 6.3.3.

Corollary 6.3.4 (Sobolev Inequality and Embedding) *Let $\varphi \in \Phi_w(\Omega)$ satisfy* (A0), (A1), (A2) *and* (aDec)$_q$ *for some $q < n$. Suppose that $\psi \in \Phi_w(\Omega)$ satisfies $t^{-\frac{1}{n}} \varphi^{-1}(x, t) \approx \psi^{-1}(x, t)$. Then for all $u \in H_0^{1,\varphi}(\Omega)$*

$$\|u\|_\psi \lesssim \|\nabla u\|_\varphi, \quad \text{and, moreover,} \quad H_0^{1,\varphi}(\Omega) \hookrightarrow L^\psi(\Omega).$$

Proof We choose $g := |\nabla f|$, $p := 1$ and $q := n'$ in Theorem 5.3.1. Then $\gamma = \frac{1}{n}$. The claim follows from Theorem 5.3.1 once we show that $M : L^{\psi_q^*}(\Omega) \to L^{\psi_q^*}(\Omega)$ is bounded. This can be done similarly as in the proof of Theorem 5.3.2. □

For Sobolev functions, convolutions approximate the identity in a quantitative way (compare with Theorem 4.4.7):

Lemma 6.3.5 *Let $\varphi \in \Phi_w(\mathbb{R}^n)$ satisfy* (A0), (A1), (A2) *and let σ be the standard mollifier. Then*

$$\|\sigma_\varepsilon * u - u\|_{L^\varphi(\mathbb{R}^n)} \lesssim \varepsilon \|\nabla u\|_{L^\varphi(\mathbb{R}^n)}$$

for every $u \in W^{1,\varphi}(\mathbb{R}^n)$. Here $\sigma_\varepsilon(t) := \frac{1}{\varepsilon^n} \sigma(\frac{t}{\varepsilon})$.

Proof Let $u \in C_0^\infty(\mathbb{R}^n)$. Using properties of the mollifier, Fubini's theorem and a change of variables, we deduce

$$u * \sigma_\varepsilon(x) - u(x) = \int_{\mathbb{R}^n} \int_0^1 \sigma_\varepsilon(y) \nabla u(x - ty) \cdot y \, dt \, dy$$

$$= \int_0^1 \int_{\mathbb{R}^n} \sigma_{\varepsilon t}(y) \nabla u(x - y) \cdot \frac{y}{t} \, dy \, dt.$$

When $\sigma_{\varepsilon t}(y) \neq 0$, $|\frac{y}{t}| \leqslant \varepsilon$. This yields the pointwise estimate

$$\left| u * \sigma_\varepsilon(x) - u(x) \right| \leqslant \varepsilon \int_0^1 \int_{\mathbb{R}^n} |\sigma_{\varepsilon t}(y)| \, |\nabla u(x - y)| \, dy \, dt$$

$$= \varepsilon \int_0^1 |\nabla u| * |\sigma_{\varepsilon t}|(x) \, dt$$

for all $u \in C_0^\infty(\mathbb{R}^n)$. Since $\operatorname{spt} \sigma_{t\varepsilon} \subset B(0, \varepsilon)$ for all $t \in [0, 1]$, the estimate is of a local character. Due to the density of $C_0^\infty(\mathbb{R}^n)$ in $W_{\text{loc}}^{1,1}(\mathbb{R}^n)$ the same estimate holds almost everywhere for all $u \in W_{\text{loc}}^{1,1}(\mathbb{R}^n)$. Hence, it holds in particular for all $u \in W^{1,\varphi}(\mathbb{R}^n)$ by Lemma 6.1.6. The pointwise estimate thus yields the norm inequality

$$\|u * \sigma_\varepsilon - u\|_\varphi \leqslant \varepsilon \left\| \int_0^1 |\nabla u| * |\sigma_{\varepsilon t}| \, dt \right\|_\varphi \leqslant Q\varepsilon \int_0^1 \big\| |\nabla u| * |\sigma_{\varepsilon t}| \big\|_\varphi \, dt$$

where we used also Corollary 3.2.5 for the Riemann sums corresponding to the integrals. By Lemma 4.4.6, we have $\big\| |\nabla u| * |\sigma_{\varepsilon t}| \big\|_\varphi \leqslant K \|\sigma\|_1 \|\nabla u\|_\varphi$. Now, the claim follows due to $\|\sigma\|_1 = 1$.   □

To prove our main compact embedding theorem, we first give a preliminary result. Since the convolution of the previous lemma "leaks" outside the domain, we need to assume that $\varphi \in \Phi_{\mathrm{w}}(\mathbb{R}^n)$, or at least $\varphi \in \Phi_{\mathrm{w}}(U)$ for some $\Omega \subset\subset U$.

Question 6.3.6 Does the next result hold assuming $\varphi \in \Phi_{\mathrm{w}}(\Omega)$ instead of $\varphi \in \Phi_{\mathrm{w}}(\mathbb{R}^n)$?

Theorem 6.3.7 *Let $\varphi \in \Phi_{\mathrm{w}}(\mathbb{R}^n)$ satisfy* (A0), (A1) *and* (aDec) *and let Ω be bounded. Then*

$$H_0^{1,\varphi}(\Omega) \hookrightarrow\hookrightarrow L^\varphi(\Omega).$$

Proof Choose a ball $B \supset \Omega$. It suffices to consider $\varphi|_{2B}$ since all convolutions in the proof are supported in this set for all small $\varepsilon > 0$. By Lemma 4.2.3, this Φ-function satisfies (A2).

Let $u_k \rightharpoonup u$ in $H_0^{1,\varphi}(\Omega)$. We write $f_k := u_k - u$ so that $f_k \rightharpoonup 0$ in $H_0^{1,\varphi}(\Omega)$. Then $\|f_k\|_{1,\varphi}$ is uniformly bounded. For every $f_k \in H_0^{1,\varphi}(\Omega)$ we choose $v_k \in C_0^\infty(\Omega)$ such that $\|f_k - v_k\|_{1,\varphi} \leqslant \frac{1}{k}$; then $v_k \rightharpoonup 0$ and $\|v_k\|_{1,\varphi}$ is uniformly bounded. By Lemma 6.1.5, $\|v_k\|_\varphi$ and $\|\nabla v_k\|_\varphi$ are uniformly bounded.

We first show that $v_k \to 0$ in $L^\varphi(\Omega)$. Let σ_ε be the standard mollifier. Then $v_k = (v_k - \sigma_\varepsilon * v_k) + \sigma_\varepsilon * v_k$ and Lemma 6.3.5 implies that

$$\|v_k\|_\varphi \lesssim \|v_k - v_k * \sigma_\varepsilon\|_\varphi + \|v_k * \sigma_\varepsilon\|_\varphi \lesssim \varepsilon \|\nabla v_k\|_\varphi + \|v_k * \sigma_\varepsilon\|_\varphi. \tag{6.3.1}$$

Since $v_k \rightharpoonup 0$ and $\varepsilon > 0$ is fixed we obtain

$$v_k * \sigma_\varepsilon(x) = \int_{\mathbb{R}^n} \sigma_\varepsilon(x - y) v_k(y) \, dy \to 0$$

as $k \to \infty$. Let $\Omega_\varepsilon := \{x \in \mathbb{R}^n : \mathrm{dist}(x, \Omega) \leqslant \varepsilon\}$. Then $v_k * \sigma_\varepsilon(x) = 0$ for all $x \in \mathbb{R}^n \setminus \Omega_\varepsilon$. By Hölder's inequality we obtain for all $x \in \Omega_\varepsilon$ that

$$|v_k * \sigma_\varepsilon(x)| = \left| \int_{\mathbb{R}^n} \sigma_\varepsilon(x - y) v_k(y) \, dy \right| \leqslant c \|v_k\|_\varphi \|\sigma_\varepsilon(x - \cdot)\|_{\varphi^*}.$$

Since $\sigma \in C_0^\infty(\mathbb{R}^n)$ we have $|\sigma| \leqslant c$ and thus $|\sigma_\varepsilon| \leqslant c\varepsilon^{-n}$. This yields $\|\sigma_\varepsilon(x - \cdot)\|_{\varphi^*} \leqslant c\varepsilon^{-n}\|\chi_{\Omega_\varepsilon}\|_{\varphi^*} \leqslant c\varepsilon^{-n}$ independently of $x \in \mathbb{R}^n$ and k. Using the uniform boundedness of v_k in L^φ we have all together proved

$$\left| v_k * \sigma_\varepsilon(x) \right| \leqslant c\varepsilon^{-n} \chi_{\Omega_\varepsilon}(x)$$

for all $x \in \mathbb{R}^n$. Since Ω is bounded we obtain by Lemma 3.7.7 that $\chi_{\Omega_\varepsilon} \in L^\varphi(\mathbb{R}^n)$. Since $v_k * \sigma_\varepsilon(x) \to 0$ almost everywhere, we obtain by dominated convergence (Lemma 3.1.4(c) and Corollary 3.3.4) that $v_k * \sigma_\varepsilon \to 0$ in $L^\varphi(\mathbb{R}^n)$ as $k \to \infty$. Hence it follows from (6.3.1) that

$$\limsup_{k\to\infty} \|v_k\|_\varphi \leqslant c\varepsilon \limsup_{k\to\infty} \|\nabla v_k\|_\varphi.$$

Since $\varepsilon > 0$ was arbitrary and $\|\nabla v_k\|_\varphi$ is uniformly bounded this yields that $v_k \to 0$ in $L^\varphi(\mathbb{R}^n)$. Hence

$$\|f_k\|_\varphi \lesssim \|f_k - v_k\|_\varphi + \|v_k\|_\varphi = \frac{c}{k} + \|v_k\|_\varphi$$

and hence $f_k \to 0$ in $L^\varphi(\mathbb{R}^n)$, i.e. $u_k \to u$ in $L^\varphi(\Omega)$ as required. $\qquad\square$

When we use Corollary 5.5.4 to interpolate between the inequalities in Corollary 6.3.4 and Theorem 6.3.7, we get the Rellich–Kondrachov Theorem for the generalized Orlicz–Sobolev spaces. The analogous result for variable Sobolev spaces was proved in [34, Corollary 8.4.4].

Theorem 6.3.8 (Compact Sobolev Embedding) *Let* $\varphi \in \Phi_w(\mathbb{R}^n)$ *satisfy* (A0), (A1) *and* (aDec)$_q$ *for* $q < n$ *and let* $\Omega \subset \mathbb{R}^n$ *be bounded domain. Suppose that* $\psi \in \Phi_w(\mathbb{R}^n)$ *satisfies* $t^{-\alpha}\varphi^{-1}(x, t) \approx \psi^{-1}(x, t)$, *for some* $\alpha \in [0, \frac{1}{n})$. *Then*

$$H_0^{1,\varphi}(\Omega) \hookrightarrow\hookrightarrow L^\psi(\Omega).$$

Remark 6.3.9 Note that for given φ in the previous theorem, there exists a suitable $\psi \in \Phi_w(\mathbb{R}^n)$ by Lemma 5.2.3. In Theorems 6.3.7 and 6.3.8 we have assumed that $\varphi \in \Phi_w(\mathbb{R}^n)$ even though the results hold in domain Ω.

Next we study modular inequalities and for that we prove a Jensen type inequality with singular measure.

Lemma 6.3.10 *Let* $B \subset \mathbb{R}^n$ *be a ball or a cube, and let* $s \geqslant 1$. *Let* $\varphi \in \Phi_w(B)$ *satisfy* (A0) *and* (A1). *Then there exists* $\beta > 0$ *such that*

$$\varphi\left(x, \beta \int_B \frac{|f(y)|}{\operatorname{diam} B\, |x - y|^{n-1}}\, dy\right) \leqslant \int_B \frac{\varphi(y, |f(y)|) + \chi_{\{f \neq 0\}}(y)}{\operatorname{diam} B\, |x - y|^{n-1}}\, dy$$

for almost every $x \in B$ *and every* $f \in L^\varphi(B)$ *with* $\varrho_\varphi(f) \leqslant 1$.

Proof Let us first note that since B is bounded, (A2) holds by Lemma 4.2.3. Hence φ satisfies (A0), (A1) and (A2).

We may assume without loss of generality that $f \geqslant 0$. Fix $r > 0$ and let $B = B(x, r)$. Define annuli $A_k := \{y \in B : 2^{-k}r \leqslant |x - y| < 2^{1-k}r\}$ for $k \geqslant 0$. We cover B with the annuli A_k and obtain

$$\int_B \frac{f(y)}{2r\,|x - y|^{n-1}}\,dy \lesssim \sum_{k=0}^{\infty} 2^{-k} \fint_{B(x,2^{1-k}r)} \chi_{A_k}(y) f(y)\,dy.$$

Let $\beta' > 0$ be from the key estimate (Theorem 4.3.2). Since $\frac{1}{2}\sum_{k=0}^{\infty} 2^{-k} = 1$, it follows by quasi-convexity (Corollary 2.2.2) that

$$(I) := \varphi\left(x, \beta\beta' \sum_{k=0}^{\infty} 2^{-k-1} \fint_{B(x,2^{1-k}r)} \chi_{A_k} f\,dy\right)$$

$$\leqslant \sum_{k=0}^{\infty} 2^{-k-1}\varphi\left(x, \beta' \fint_{B(x,2^{1-k}r)} \chi_{A_k} f\,dy\right).$$

Next we apply the key estimate (Theorem 4.3.2) with $p = 1$. Thus for some $h \in L^1(B) \cap L^\infty(B)$ we have

$$\varphi\left(x, \beta' \fint_{B(x,2^{1-k}r)} \chi_{A_k} f\,dy\right) \leqslant \fint_{B(x,2^{1-k}r)} \varphi(y, \chi_{A_k} f)\,dy$$

$$+ \|h\|_\infty \frac{\left|\{\chi_{A_k} f \neq 0\} \cap B(x, 2^{1-k}r) \cap B\right|}{|B(x, 2^{1-k}r)|}$$

$$\leqslant \frac{1 + \|h\|_\infty}{(2^{-k}r)^n} \int_{A_k} \varphi(y, f) + \chi_{\{f \neq 0\}}(y)\,dy.$$

Combining the two estimates, we find that

$$(I) \lesssim \sum_{k=1}^{\infty} \frac{1}{(2^{-k}r)^n} \int_{A_k} \varphi(y, f) + \chi_{\{f \neq 0\}}(y)\,dy \approx \int_B \frac{\varphi(y, f) + \chi_{\{f \neq 0\}}(y)}{2r\,|x - y|^{n-1}}\,dy.$$

By (aInc)$_1$ this is the claim for suitable β_2. $\square$

The next proposition is a Sobolev–Poincaré inequality for weak Φ-functions and yields an exponent strictly less than 1. For example this is the main requirement for Gehring's lemma. The proof introduces a probability measure that allows Jensen's inequality to be used in the usual setting. This technique was used in [36]. The rest of the proof consists of handling left-over terms and technicalities.

Question 6.3.11 Does the following result hold with $s = n'$?

Proposition 6.3.12 *Let $B \subset \mathbb{R}^n$ be a ball or a cube, $s \in [1, n')$ and $\varphi^{\frac{1}{s}} \in \Phi_w(B)$ satisfy assumptions (A0) and (A1). Then there exists a constant $\beta \in (0, 1]$ such that the following hold.*

(a) *For every $u \in H_0^{1,1}(B)$ with $\varrho_{\varphi^{\frac{1}{s}}}(|\nabla u|) \leqslant 1$ we have*

$$\fint_B \varphi\left(x, \beta\frac{|u|}{\operatorname{diam} B}\right) dx \leqslant \left(\fint_B \varphi(x, |\nabla u|)^{\frac{1}{s}} dx\right)^s + \frac{\left|\{|\nabla u| \neq 0\} \cap B\right|}{|B|}.$$

(b) *For every $v \in W^{1,1}(B)$ with $\varrho_{\varphi^{\frac{1}{s}}}(|\nabla u|) \leqslant 1$ we have*

$$\fint_B \varphi\left(x, \beta\frac{|u - u_B|}{\operatorname{diam} B}\right) dx \leqslant \left(\fint_B \varphi(x, |\nabla u|)^{\frac{1}{s}} dx\right)^s + \frac{\left|\{|\nabla u| \neq 0\} \cap B\right|}{|B|}.$$

Proof Note first that since $\varphi^{\frac{1}{s}} \in \Phi_w(B)$ satisfies (A0), we have $\varphi \in \Phi_w(B)$ satisfying (A0). The proofs of the claims are similar and hence we only prove (b). For brevity, we denote $\kappa := \operatorname{diam} B$.

By Lemma 6.2.3(b) (or more precisely a corresponding result in a convex domain [46, Lemma 7.16, p. 162]) we have, for almost every $x \in B$,

$$|u(x) - u_B| \lesssim \int_B \frac{|\nabla u(y)|}{|x - y|^{n-1}} dy.$$

Let us write $f(y) := \varphi(y, |\nabla u|)^{1/s} + \chi_{\{|\nabla u| \neq 0\}}(y)$. The previous inequality and Lemma 6.3.10 with function $\varphi^{\frac{1}{s}}$ and constant $\beta' = \frac{\beta}{c_1}$ yield

$$\varphi\left(x, \beta'\frac{|u(x) - u_B|}{\kappa}\right) \leqslant \varphi\left(x, \beta\int_B \frac{|\nabla u(y)|}{\kappa|x - y|^{n-1}} dy\right)$$
$$\leqslant \left(\int_B \frac{f(y)}{\kappa|x - y|^{n-1}} dy\right)^s. \tag{6.3.2}$$

Set $J := \int_B f(y)\,dy$. If $J = 0$, then $|\nabla u| = 0$ almost everywhere and u is a constant function. Thus $u = u_B$ almost everywhere and hence the claim follows. If $J > 0$, then we define a probability measure by $d\mu(y) := \frac{1}{J}f(y)\,dy$. Since μ is a probability measure, we can use Jensen's inequality for the convex function $t \mapsto t^s$:

$$\left(\int_B \frac{f(y)}{\kappa|x - y|^{n-1}} dy\right)^s \leqslant \int_B \frac{J^s}{\kappa^s|x - y|^{s(n-1)}} d\mu(y)$$
$$= \frac{J^{s-1}}{\kappa^s}\int_B \frac{f(y)}{|x - y|^{s(n-1)}} dy.$$

We integrate the previous inequality over $x \in B$, and use Fubini's theorem to change the order of integration

$$\fint_B \left(\int_B \frac{f(y)}{\kappa |x - y|^{n-1}} \, dy \right)^s dx \leqslant \frac{J^{s-1}}{\kappa^s} \fint_B f(y) \int_B \frac{dx}{|x - y|^{s(n-1)}} \, dy.$$

Finally, we use the assumption $s < \frac{n}{n-1}$ and $\kappa = \operatorname{diam} B$ to estimate

$$\int_B \frac{dx}{|x - y|^{s(n-1)}} \leqslant c(n)\kappa^{-s(n-1)+n}$$

for $y \in B$. We conclude, taking into account the definition of J, that

$$\fint_B \left(\int_B \frac{f(y)}{\kappa |x - y|^{n-1}} \, dy \right)^s dx \lesssim \left(\frac{J}{\kappa^n} \right)^s$$

$$= \left(\fint_B \varphi(y, |\nabla u|)^{\frac{1}{s}} + \chi_{\{|\nabla u| \neq 0\}}(y) \, dy \right)^s$$

$$\approx \left(\fint_B \varphi(y, |\nabla u|)^{\frac{1}{s}} \, dy \right)^s + \frac{|\{|\nabla u| \neq 0\} \cap B|}{|B|}$$

Combining this with (6.3.2) integrated over B, we complete the proof as the constant on the right-hand side can be absorbed into β' by (aInc)$_1$.   □

Corollary 6.3.13 (Sobolev–Poincaré Inequality in Modular Form) *Let $B \subset \mathbb{R}^n$ be a ball or a cube, and $\varphi \in \Phi_w(B)$ satisfy assumptions* (A0) *and* (A1), *and let $1 \leqslant s < n'$. Then there exists a constant $\beta \in (0, 1]$ such that the following hold.*

(a) *For every $u \in H_0^{1,1}(B)$ with $\varrho_\varphi(|\nabla u|) \leqslant 1$ we have*

$$\left(\fint_B \varphi\left(x, \beta \frac{|u|}{\operatorname{diam} B} \right)^s dx \right)^{\frac{1}{s}} \leqslant \fint_B \varphi(x, |\nabla u|) \, dx + \frac{|\{|\nabla u| \neq 0\} \cap B|}{|B|}.$$

(b) *For every $v \in W^{1,1}(B)$ with $\varrho_\varphi(|\nabla u|) \leqslant 1$ we have*

$$\left(\fint_B \varphi\left(x, \beta \frac{|u - u_B|}{\operatorname{diam} B} \right)^s dx \right)^{\frac{1}{s}} \leqslant \fint_B \varphi(x, |\nabla u|) \, dx + \frac{|\{|\nabla u| \neq 0\} \cap B|}{|B|}.$$

Proof Let us write $\psi := \varphi^s$. Then $\psi^{\frac{1}{s}}$ satisfies the assumptions of Proposition 6.3.12. By the assumption we have

$$\int_B \psi(x, |\nabla u|)^{\frac{1}{s}} \, dx = \int_B \varphi(x, |\nabla u|) \, dx \leqslant 1.$$

Hence we can apply Proposition 6.3.12 to $\psi^{\frac{1}{s}}$ and obtain by (a) that

$$\fint_B \psi\left(x, \beta \frac{|u|}{\operatorname{diam} B}\right) dx \leqslant \left(\fint_B \psi(x, |\nabla u|)^{\frac{1}{s}} dx + \frac{|\{|\nabla u| \neq 0\} \cap B|}{|B|}\right)^s.$$

This yields the claim when we raise both sides to power $\frac{1}{s}$. The case (b) follows similarly. $\qquad\square$

6.4 Density of Regular Functions

This section contains a sufficient condition for when every function in a Sobolev space can be approximated by a more regular function, such as a smooth or Lipschitz continuous function.

Lemma 6.4.1 *Let $\varphi \in \Phi_w(\Omega)$ satisfy (aDec). Then Sobolev functions with bounded support in $\mathbb{R}^n$ are dense in $W^{1,\varphi}(\Omega)$.*

Proof Let us denote $B_r := B(0, r)$, $r \geqslant 1$. Let $\eta_r \in C_0^\infty(\mathbb{R}^n)$ be a cut-off function with $\eta_r = 1$ on B_r, $\eta_r = 0$ on $\mathbb{R}^n \setminus B_{r+1}$, $0 \leqslant \eta_r \leqslant 1$ and $|\nabla \eta_r| \leqslant 2$. We show that $u\eta_r \to u$ in $W^{1,\varphi}(\Omega)$ as $r \to \infty$. Note first that

$$\|u - u\eta_r\|_{W^{1,\varphi}(\Omega)} \lesssim \|u\|_{W^{1,\varphi}(\Omega \setminus B_{r+1})} + \|u - u\eta_r\|_{W^{1,\varphi}(\Omega \cap (B_{r+1} \setminus B_r))}.$$

Since φ satisfies (aDec), modular convergence and norm convergence are equivalent by Corollary 3.3.4. Lemma 6.1.5 and the absolutely continuity of the integral imply that

$$\varrho_{W^{1,\varphi}(\Omega \setminus B_{r+1})}(u) \approx \varrho_{L^\varphi(\Omega \setminus B_{r+1})}(u) + \varrho_{L^\varphi(\Omega \setminus B_{r+1})}(\nabla u) \to 0$$

as $r \to \infty$ and hence also $\|u\|_{W^{1,\varphi}(\Omega \setminus B_{r+1})} \to 0$. To handle the second term in the above inequality we observe that

$$|\nabla u - \nabla(u\eta_r)| \leqslant (1 - \eta_r)|\nabla u| + |\nabla \eta_r| |u| \leqslant |\nabla u| + 2|u|.$$

Thus $\|u - u\eta_r\|_{W^{1,\varphi}(\Omega \cap (B_{r+1} \setminus B_r))} \leqslant \||\nabla u| + 4|u|\|_{L^\varphi(\Omega \setminus B_r)}$. As with the previous term, this converges to 0 when $r \to \infty$. $\qquad\square$

It has been showed in [34, Example 9.1.5] that Sobolev functions with compact support are not dense in $W^{1,\varphi}(\mathbb{R})$ when $\varphi(x, t) := t^{\max\{|x|, 1\}}$. Hence the assumption (aDec) is not redundant in Lemma 6.4.1.

Lemma 6.4.2 *Let $\varphi \in \Phi_w(\Omega)$ satisfy (aDec) and $L^\varphi \subset L_{loc}^1$. Then bounded Sobolev functions with compact support in $\mathbb{R}^n$ are dense in $W^{1,\varphi}(\Omega)$.*

Proof Let $u \in W^{1,\varphi}(\Omega)$. By Lemma 6.4.1, we may assume that u has bounded support $\Omega \cap B(0, r)$ for some large $r > 0$. We define the truncation u_m by

$$u_m(x) := \max\{\min\{u(x), m\}, -m\}$$

for $m > 0$. By Lemma 6.1.7, $u_m \in W^{1,1}_{\text{loc}}(\Omega)$. Since $|u_m| \leqslant |u|$ and $|\nabla u_m| \leqslant |\nabla u|$ we obtain by solidity (3.3.2) that $u_m \in W^{1,\varphi}(\Omega)$.

Since φ satisfies (aDec), we have by Lemma 3.2.9 that $\int_\Omega \varphi(x, u) \, dx < \infty$ so that $|\{x \in \Omega : \varphi(x, u) \geqslant m\}| \to 0$ as $m \to \infty$. If $u \geqslant am^3 \geqslant m$, then the inequality (from (aInc)₁)

$$\varphi(x, m) \leqslant \frac{a}{u} m \varphi(x, u)$$

implies that either $\varphi(x, m) \leqslant 1$ or $\varphi(x, u) \geqslant m^2$. Hence, for large m,

$$\{|u| \geqslant am^3\} \subset \left(\{\varphi(\cdot, m) \leqslant 1\} \cap \text{spt}\, u\right) \cup \{\varphi(\cdot, u) \geqslant m^2\}.$$

Since $\varphi(x, t) \to \infty$ when $t \to \infty$, we find that

$$\left|\{\varphi(\cdot, m) \leqslant 1\} \cap \text{spt}\, u\right| \to 0$$

as $m \to \infty$. Since φ satisfies (aDec), $\varphi(x, t) < \infty$ for almost every $x \in \Omega$ and all $t \in [0, \infty)$. Thus $|\{\varphi(x, u) \geqslant m^2\}| \to 0$ as $m \to \infty$.

Lemma 6.1.5 and the absolutely continuity of the integral yield

$$\varrho_{1,\varphi}\left(u - u_{am^3}\right) \lesssim \int_{\{|u| \geqslant am^3\}} \varphi(x, u) + \varphi(x, |\nabla u|) \, dx \to 0$$

as $m \to \infty$. Since φ satisfies (aDec), norm convergence follows by Corollary 3.3.4. $\qquad\square$

The next example shows that bounded functions are not always dense in the Sobolev space.

Example 6.4.3 Let $\Omega = (1, \infty) \subset \mathbb{R}$ and $\varphi(x, t) := \left(\frac{t}{x}\right)^x$. Then $\varphi \in \Phi_s(1, \infty)$ does not satisfy (aDec). Let $u(x) := x$, so that $u'(x) = 1$. Next we show that $u \in W^{1,\varphi}(1, \infty)$:

$$\int_1^\infty \varphi(x, \lambda u) \, dx = \int_1^\infty \lambda^x \, dx < \infty$$

for $\lambda \in (0, 1)$; and

$$\int_1^\infty \varphi(x, u') \, dx = \int_1^\infty \left(\frac{1}{x}\right)^x \, dx < 1.$$

Let u_m be the truncation of u defined in Lemma 6.4.2. Then

$$\int_1^\infty \varphi\left(x, \frac{u - u_m}{\lambda}\right) dx = \int_m^\infty \left(\frac{x - m}{\lambda x}\right)^x dx \geqslant \int_{2m}^\infty \left(\frac{\frac{1}{2}x}{\lambda x}\right)^x dx = \infty$$

for $\lambda \in (0, \frac{1}{2})$ and hence $\|u - u_m\|_\varphi \geqslant \frac{1}{2}$ for all $m \geqslant 1$.

Next we give short proofs for the density of smooth functions based on the boundedness of convolution. For Lebesgue spaces this has been shown in Theorem 3.7.15.

Theorem 6.4.4 *Let* $\varphi \in \Phi_w(\mathbb{R}^n)$ *satisfy* (A0), (A1), (A2) *and* (aDec). *Then* $C_0^\infty(\mathbb{R}^n)$ *is dense in* $W^{1,\varphi}(\mathbb{R}^n)$.

Proof Let $u \in W^{1,\varphi}(\mathbb{R}^n)$ and let $\varepsilon > 0$ be arbitrary. By Lemma 6.4.1, we may assume that u has compact support in $\mathbb{R}^n$. Let σ_ε be a standard mollifier. Then $u * \sigma_\varepsilon$ belongs to $C_0^\infty(\mathbb{R}^n)$ and

$$\nabla(u * \sigma_\varepsilon) - \nabla u = (\nabla u) * \sigma_\varepsilon - \nabla u.$$

Thus the claim follows since $\|f * \sigma_\varepsilon - f\|_\varphi \to 0$ as $\varepsilon \to 0^+$ for every $f \in L^\varphi(\mathbb{R}^n)$ by Theorem 4.4.7 and $\|u\|_{W^{1,\varphi}(\Omega)} \approx \|u\|_{L^\varphi(\Omega)} + \|\nabla u\|_{L^\varphi(\Omega)}$ by Lemma 6.1.5. $\quad\square$

Then we study density of smooth functions in a subset of $\mathbb{R}^n$.

Lemma 6.4.5 *Let* $\varphi \in \Phi_w(\Omega)$ *satisfy* (A0), (A1), (A2) *and* (aDec). *Let* $u \in W^{1,\varphi}(\Omega)$, $D \subset\subset \Omega$ *and* σ_ε *be the standard mollifier. Then* $\sigma_\varepsilon * u \to u$ *in* $W^{1,\varphi}(D)$ *as* $\varepsilon \to 0^+$.

Proof Let $\varepsilon < \mathrm{dist}(D, \partial\Omega)$. We use the formula $\nabla(\sigma_\varepsilon * u) = \sigma_\varepsilon * \nabla u$ in D. Then the claim follows since $\|u\|_{W^{1,\varphi}(\Omega)} \approx \|u\|_{L^\varphi(\Omega)} + \|\nabla u\|_{L^\varphi(\Omega)}$ by Lemma 6.1.5, and $\|f * \sigma_\varepsilon - f\|_\varphi \to 0$ as $\varepsilon \to 0^+$ for every $f \in L^\varphi(\Omega)$ by Theorem 4.4.7. $\quad\square$

For the next result, we need a partition of unity. A proof for the existence of a partition of unity can be found for example in [2, Theorem 3.14, p. 51] or in [88, Theorem 1.44, p. 25].

Theorem 6.4.6 (Partition of Unity) *Let* $\mathcal{U}$ *be a family of open sets which cover a closed set* $E \subset \mathbb{R}^n$. *Then there exists a family* $\mathcal{F}$ *of functions in* $C_0^\infty(\mathbb{R}^n)$ *with values in* $[0, 1]$ *such that*

(a) $\sum_{f \in \mathcal{F}} f(x) = 1$ *for every* $x \in E$;
(b) *for each function* $f \in \mathcal{F}$, *there exists* $U \in \mathcal{U}$ *such that* spt $f \subset U$; *and*
(c) *if* $K \subset E$ *is compact, then* spt $f \cap K \neq \emptyset$ *for only finitely many* $f \in \mathcal{F}$.

The family $\mathcal{F}$ *is said to be a* partition of unity *of* E *subordinate to the open covering* $\mathcal{U}$.

Theorem 6.4.7 *Let* $\varphi \in \Phi_w(\Omega)$ *satisfy* (A0), (A1), (A2) *and* (aDec). *Then* $C^\infty(\Omega) \cap W^{1,\varphi}(\Omega)$ *is dense in* $W^{1,\varphi}(\Omega)$.

Proof Let $u \in W^{1,\varphi}(\Omega)$. Fix $\varepsilon > 0$ and define $\Omega_0 := \emptyset$,

$$\Omega_m := \left\{ x \in \Omega : \mathrm{dist}(x, \partial\Omega) > \tfrac{1}{m} \right\} \quad \text{and} \quad U_m := \Omega_{m+1} \setminus \overline{\Omega}_{m-1}$$

for $m = 1, 2, \ldots$ Let (ξ_m) be a partition of unity subordinate to the covering (U_m). Let σ_δ be the standard mollifier. For every m there exists δ_m such that $\mathrm{spt}\left((\xi_m u) * \sigma_{\delta_m}\right) \subset U_m \subset\subset \Omega$ and by Lemma 6.4.5 we conclude by choosing a smaller δ_m if necessary, that $\|(\xi_m u) - (\xi_m u) * \sigma_{\delta_m}\|_{W^{1,\varphi}(\Omega)} \leqslant \varepsilon \, 2^{-m}$. We define

$$u_\varepsilon := \sum_{m=1}^{\infty} (\xi_m u) * \sigma_{\delta_m} \, .$$

Every point $x \in \Omega$ has a neighborhood such that the above sum has only finitely many non-zero terms and thus $u_\varepsilon \in C^{\infty}(\Omega)$. Furthermore, this is an approximating sequence, since by quasi-convexity (Corollary 3.2.5)

$$\|u - u_\varepsilon\|_{W^{1,\varphi}(\Omega)} \lesssim \sum_{m=1}^{\infty} \|\xi_m u - (\xi_m u) * \sigma_{\delta_m}\|_{W^{1,\varphi}(\Omega)} \leqslant \varepsilon.$$

$\square$

Chapter 7
Special Cases

In this chapter, we consider our conditions and results in some special cases, namely variable exponent spaces and their variants, for double phase and degenerate double phase growth, as well as for Orlicz growth without x-dependence. The following table summarizes the results regarding the conditions. The Φ-functions are from Example 2.5.3.

Let us mention that also some other types of Φ-functions have been considered in the literature which can also be handled as special cases (Table 7.1). For instance, Φ-functions of the type $\varphi(t)^{p(x)}$ or $\varphi(t^{p(x)})$ have been studied in [17, 44, 45] by Capone, Cruz-Uribe, Fiorenza, Giannetti, Passarelli di Napoli, Ragusa, and Tachikawa and the double-exponent case $t^{p(x)} + t^{q(x)}$ has been considered by Cencelj, Rădulescu, Repovš and Zhang [19, 119, 128].

7.1 Variable Exponent Growth

We consider the variable exponent case, $\varphi(x, t) = t^{p(x)}$ for some measurable function $p : \Omega \to [1, \infty]$. Here we interpret $t^\infty := \infty \chi_{(1,\infty]}(t)$. Then one can check that $\varphi^{-1}(x, t) = t^{1/p(x)}$, where $t^{1/\infty} := \chi_{(0,\infty)}(t)$. More information about variable exponent spaces can be found from the books [18, 27, 34, 38, 72, 73, 115].

Recently, more specialized questions have been studied in the $L^{p(\cdot)}$-case, such as quasiminimizers by Adamowicz and Toivanen [1], minimizers when $p^- = 1$ or $p^+ = \infty$ by Karagiorgos and Yannakakis [67, 68] and spaces of analytical functions by Karapetyants and Samko [70]. Note that also variants such as

$$\varphi(x, t) = t^{p(x)} \log(e + t)^{q(x)}$$

© Springer Nature Switzerland AG 2019
P. Harjulehto, P. Hästö, *Orlicz Spaces and Generalized Orlicz Spaces*,
Lecture Notes in Mathematics 2236,
https://doi.org/10.1007/978-3-030-15100-3_7

Table 7.1 Conditions in variable exponent and double phase cases

$\varphi(x,t)$	(A0)	(A1)	(A2)	(aInc)	(aDec)
$t^{p(x)}a(x)$	$a \approx 1$	$\frac{1}{p} \in C^{\log}$	Nekvinda	$p^- > 1$	$p^+ < \infty$
$t^{p(x)} \log(e+t)$	True	$\frac{1}{p} \in C^{\log}$	Nekvinda	$p^- > 1$	$p^+ < \infty$
$t^p + a(x)t^q$	$a \in L^\infty$	$a \in C^{0,\frac{n}{p}(q-p)}$	True	$p > 1$	$q < \infty$
$(t-1)_+^p + a(x)(t-1)_+^q$	$a \in L^\infty$	$a \in C^{0,\frac{n}{p}(q-p)}$	True	$p > 1$	False

can be handled in our framework. These variants has been studied for example in [43, 47, 92, 93, 95, 108–110] by Giannetti, Giova, Mizuta, Nakai, Ohno, Ok, Passarelli di Napoli, and Shimomura.

Recall some well-known concepts from variable exponent spaces. We define

$$p^- := \operatorname*{ess\,inf}_{x \in \Omega} p(x) \quad \text{and} \quad p^+ := \operatorname*{ess\,sup}_{x \in \Omega} p(x).$$

We say that $\frac{1}{p}$ is log-*Hölder continuous*, $\frac{1}{p} \in C^{\log}$, if

$$\left| \frac{1}{p(x)} - \frac{1}{p(y)} \right| \leqslant \frac{c}{\log(e + \frac{1}{|x-y|})}$$

for every distinct $x, y \in \Omega$. This condition was first used in the variable exponent spaces by Zhikov [129]. Our formulation is from Section 4.1 of [34].

Like in the generalized Orlicz case, for variable exponents one needs a local condition and a decay condition. Nekvinda's decay condition for variable exponent spaces can be stated as follows: $1 \in L^{s(\cdot)}(\Omega)$, with $\frac{1}{s(x)} = |\frac{1}{p(x)} - \frac{1}{p_\infty}|$ and $p_\infty \in [1, \infty]$. Equivalently, this means that there exists $c > 0$ such that

$$\int_{\{p(x) \neq p_\infty\}} c^{\frac{1}{|\frac{1}{p(x)} - \frac{1}{p_\infty}|}} \, dx < \infty.$$

This condition was originally introduced by Nekvinda in [100]. We observe that the log-Hölder decay condition,

$$\left| \frac{1}{p(x)} - \frac{1}{p_\infty} \right| \leqslant \frac{c}{\log(e + |x|)},$$

implies Nekvinda's decay condition. This and the log-Hölder decay condition are discussed in [34, Section 4.2]. We now connect these conditions to our framework.

Lemma 7.1.1 *The Φ-function $\varphi(x,t) = t^{p(x)}$ satisfies* (A0), (Inc)$_{p^-}$ *and* (Dec)$_{p^+}$.

Proof Since $\varphi^{-1}(x, 1) = 1$, (A0) holds. Since $p(x) \geqslant p^-$, we obtain for $0 \leqslant s \leqslant t$ that

$$\frac{\varphi(x, s)}{s^{p^-}} = s^{p(x)-p^-} \leqslant t^{p(x)-p^-} = \frac{\varphi(x, t)}{t^{p^-}}$$

for almost every x and so (Inc)$_{p^-}$ holds. The proof for (Dec)$_{p^+}$ is similar. □

Proposition 7.1.2 *The Φ-function $\varphi(x, t) = t^{p(x)}$ satisfies (A1) if and only if $\frac{1}{p}$ is log-Hölder continuous.*

Proof Let $x, y \in \Omega$ satisfy $|x - y| \leqslant \frac{1}{2\omega_n^{1/n}}$, where ω_n is the measure of the unit ball. Let $B \subset \mathbb{R}^n$ be such a ball that $x, y \in B \cap \Omega$ and $\text{diam}(B) = 2|x - y|$, and note that $|B| \leqslant 1$. By symmetry, we may assume that $p(x) < p(y)$. Since $\varphi^{-1}(x, t) = t^{1/p(x)}$, it follows from assumption (A1) that $\beta t^{1/p(x)} \leqslant t^{1/p(y)}$ for $t \in [1, \frac{1}{|B|}]$. We observe that $(\omega_n |x - y|^n)^{-1} = \frac{1}{|B|}$ and so we conclude that

$$(\omega_n^{-1} |x - y|^{-n})^{\frac{1}{p(x)} - \frac{1}{p(y)}} \leqslant \frac{1}{\beta}.$$

Taking the logarithm, we find that

$$\frac{1}{p(x)} - \frac{1}{p(y)} \leqslant \frac{\log \frac{1}{\beta}}{n \log(|x - y|^{-1}) - \log(\omega_n)} \lesssim \frac{1}{\log(e + |x - y|^{-1})}.$$

If $|x - y| > \frac{1}{2\omega_n^{1/n}}$, then the previous inequality holds trivially, since $\frac{1}{p} \in [0, 1]$. Hence we have shown that the condition (A1) implies that $\frac{1}{p}$ is log-Hölder continuous.

For the opposite implication, assume that B is a ball with $|B| \leqslant 1$ and $x, y \in B \cap \Omega$. By symmetry, we may assume that $p(x) < p(y)$. If $t \in [1, \frac{1}{|B|}]$, then

$$\frac{\varphi^{-1}(x, t)}{\varphi^{-1}(y, t)} = t^{\frac{1}{p(x)} - \frac{1}{p(y)}} \leqslant |B|^{-\frac{c}{\log(e + \frac{1}{|x-y|})}} \leqslant |B(x, \tfrac{1}{2}|x - y|)|^{-\frac{c}{\log(e + \frac{1}{|x-y|})}}$$

$$= e^{\frac{cn \log \frac{1}{|x-y|}}{\log(e + \frac{1}{|x-y|})}} \leqslant e^{cn} < \infty.$$

This yields that $\varphi^{-1}(x, t) \lesssim \varphi^{-1}(y, t)$ and hence (A1) follows. □

We consider next the decay of the exponent at infinity and link it to (A2).

Proposition 7.1.3 *The Φ-function $\varphi(x, t) = t^{p(x)}$ satisfies (A2) if and only if p satisfies Nekvinda's decay condition.*

Proof By Lemma 4.2.7, it suffices to check (A2').

Let $\alpha, \beta \in (0, 1)$. By considering the zero of the derivative (at the value $s = (\alpha\beta^\alpha)^{1/(1-\alpha)}$), we conclude that

$$\max_{s\in[0,1]} \left((\beta s)^\alpha - s\right) = \beta^{\frac{\alpha}{1-\alpha}} \alpha^{\frac{\alpha}{1-\alpha}} (1 - \alpha).$$

Since $\frac{\alpha \log \alpha}{1-\alpha}$ is bounded on $(0, 1)$, we find that $\alpha^{\frac{\alpha}{1-\alpha}} \approx 1$. Therefore,

$$\max_{s\in[0,1]} \left((\beta s)^\alpha - s\right) \approx \beta^{\frac{\alpha}{1-\alpha}} (1 - \alpha). \tag{7.1.1}$$

Suppose first that φ satisfies (A2′). By Lemma 4.2.9 we may assume that $s = 1$ in (A2′). Choose a sequence $x_i \to \infty$ with $h(x_i) \to 0$. Let us write $p_\infty := \limsup_{i\to\infty} p(x_i)$. Assume that $p(x)$, $p_\infty < \infty$ (with standard modification if one of the exponents is infinite). By taking limit inferior of both inequalities in (A2′) we obtain for $t \in [0, 1]$ that

$$(\beta t)^{p_\infty} = \liminf_{i\to\infty} \varphi(x_i, \beta t) \leqslant \varphi_\infty(t) + \lim_{i\to\infty} h(x_i) = \varphi_\infty(t)$$

$$\varphi_\infty(\beta t) \leqslant \liminf_{i\to\infty}(\varphi(x_i, t) + h(x_i)) = t^{p_\infty}.$$

Thus $\varphi_\infty(t) \simeq t^{p_\infty}$ for $t \in [0, 1]$ and by (A2′) there exist $\beta > 0$ and $h \in L^1(\Omega) \cap L^\infty(\Omega)$, such that

$$\begin{cases} (\beta t)^{p(x)} \leqslant t^{p_\infty} + h(x), \\ (\beta t)^{p_\infty} \leqslant t^{p(x)} + h(x). \end{cases}$$

If $p(x) < p_\infty$, then (7.1.1), with $s := t^{p_\infty}$, $\alpha := \frac{p(x)}{p_\infty}$ and $\beta := \beta^{p_\infty}$, implies that

$$(\beta^{p_\infty})^{\frac{p(x)}{p_\infty - p(x)}} \frac{p_\infty - p(x)}{p_\infty} \approx \max_{t\in[0,1]} \left((\beta t)^{p(x)} - t^{p_\infty}\right) \leqslant h(x).$$

The other inequality gives a similar expression when $p(x) > p_\infty$. Since $h \in L^1(\Omega)$, these imply that

$$\beta^{\frac{1}{\left|\frac{1}{p_\infty} - \frac{1}{p(\cdot)}\right|}} \frac{|p_\infty - p(\cdot)|}{\max\{p_\infty, p(\cdot)\}} \in L^1(\Omega).$$

Since the second factor can be bounded by a power of the first, this implies Nekvinda's condition.

Assume now conversely that Nekvinda's decay condition holds. Then there exist $c_1 \in (0, 1)$ and $p_\infty \in [1, \infty]$ such that

$$\int_{\{p(x) \neq p_\infty\}} c_1^{\frac{1}{\left|\frac{1}{p_\infty} - \frac{1}{p(x)}\right|}} \, dx < \infty.$$

It follows from Young's inequality that

$$(c_1 s)^{p(x)} \leqslant s^{p_\infty} + c_1^{\frac{1}{\frac{1}{p_\infty} - \frac{1}{p(x)}}}$$

when $p(x) < p_\infty$ and analogously in the opposite case. Hence (A2$'$) holds by Lemma 4.2.9 with

$$h(x) := c_1^{\frac{1}{\left|\frac{1}{p_\infty} - \frac{1}{p(x)}\right|}}.$$

$\square$

Combining these propositions with the other results from this book, we obtain numerous results for variable exponent spaces. In particular, by Lemma 7.1.1, Proposition 7.1.2, Proposition 7.1.3 and Theorem 4.3.4 we obtain the following corollary.

Corollary 7.1.4 *Let* $\Omega \subset \mathbb{R}^n$ *and* $\varphi(x, t) := t^{p(x)}$. *If* $p^- > 1$, $\frac{1}{p}$ *is log-Hölder continuous and satisfies Nekvinda's decays condition, then the maximal operator is bounded on* $L^{p(\cdot)}(\Omega)$.

7.2 Double Phase Growth

In this section we consider several double phase functionals, namely:

$$t^p + a(x)t^q, \quad t^p + a(x)t^p \log(e + t) \quad \text{and} \quad (t - 1)_+^p + a(x)(t - 1)_+^q \quad (7.2.1)$$

with $1 \leqslant p < q < \infty$ and $a \in L^\infty(\Omega)$ non-negative. The basic variant with continuous a was considered by Zhikov [129] and has been recently studied by Baroni, Colombo and Mingione [7–9, 25, 26] as well as other researchers, e.g. Byun, Cho, Colasuonno, Oh, Ok, Perera, Ryu, Shin, Squassina and Youn [13–16, 24, 111, 112, 114]. The case when a attains only values 0 and 1, has been studied by Barroso and Zappale [10].

Most of our conditions are easily checked for these functionals and are collected in the next result.

Proposition 7.2.1 *Let* $\varphi \in \Phi_c(\Omega)$, *be one of the functionals in* (7.2.1). *Then* φ *satisfies* (A0), (A2) *and* (Inc)$_p$. *The first two satisfy* (aDec).

Proof All three functionals satisfy the condition of Corollary 3.7.5 since $1 \leqslant \varphi(x, 2) \leqslant 2^p + 2^q \|a\|_\infty$, so (A0) holds.

When $t \in [0, s]$, $t^p + a(x)t^q \approx t^p + a(x)t^p \log(e + t) \approx t^p$, where the constants depend on s, so (A2) follows from Lemma 4.2.5. For the same range, $(t - 1)_+^p + a(x)(t - 1)_+^q \approx (t - 1)_+^p$, so (A2) is clear in this case, as well.

The claims regarding (Inc)$_p$ and (aDec) of follow since t^r is increasing for $r > 0$, more precisely, the first function satisfies (aDec)$_q$ and the second satisfies (aDec)$_{p+\varepsilon}$ for any $\varepsilon > 0$. $\square$

It remains to investigate (A1).

Proposition 7.2.2 *The* Φ*-function* $\varphi(x, t) = t^p + a(x)t^q$ *satisfies* (A1) *if and only if*

$$a(y) \lesssim a(x) + |x - y|^{\frac{n}{p}(q-p)}$$

for every $x, y \in \Omega$.

In particular, if $a \in C^{\frac{n}{p}(q-p)}(\Omega)$, *then* φ *satisfies* (A1).

Proof We observe that $\varphi(x, t) \approx \max\{t^p, a(x)t^q\}$. Hence we conclude that $\varphi^{-1}(x, t) \approx \min\{t^{1/p}, (t/a(x))^{1/q}\}$ and the (A1) condition becomes, after we divide by $t^{1/q}$,

$$\beta \min\{t^{\frac{1}{p} - \frac{1}{q}}, a(x)^{-\frac{1}{q}}\} \leqslant \min\{t^{\frac{1}{p} - \frac{1}{q}}, a(y)^{-\frac{1}{q}}\}$$

for $x, y \in B \cap \Omega$, $|B| \leqslant 1$ and $t \in [1, \frac{1}{|B|}]$. We may assume that $\operatorname{diam}(B) \leqslant 2|x - y|$. The case $a(y)^{-1/q} \geqslant t^{1/p - 1/q}$ is trivial, so the condition is equivalent to

$$\beta \min\{t^{\frac{1}{p} - \frac{1}{q}}, a(x)^{-\frac{1}{q}}\} \leqslant a(y)^{-\frac{1}{q}}$$

for $a(y)^{-1/q} < t^{1/p - 1/q}$. Since the exponent of t is positive, we need only check the inequality for the upper bound of t, namely $t = \frac{1}{|B|}$. Furthermore, $|B| \approx |x - y|^n$. Thus the condition is further equivalent to

$$\beta' \min\{|x - y|^{-n(\frac{1}{p} - \frac{1}{q})}, a(x)^{-\frac{1}{q}}\} \leqslant a(y)^{-\frac{1}{q}}.$$

Raising both sides to the power $-q$, we obtain the equivalent condition

$$a(y) \lesssim \max\{|x - y|^{\frac{n}{p}(q-p)}, a(x)\} \approx |x - y|^{\frac{n}{p}(q-p)} + a(x).$$

If $a \in C^{\frac{n}{p}(q-p)}(\Omega)$, then this certainly holds. $\square$

The previous result also holds for the degenerate double phase functional $(t - 1)_+^p + a(x)(t - 1)_+^q$, since only large values of t are important for (A1) and this functional is independent of t when $t \in [0, 1]$.

Combining these propositions with the other results from this book, we obtain numerous results for the double phase functional and its relatives, many of them new. In particular, by Propositions 7.2.1 and 7.2.2 and Theorem 4.3.4 we obtain the following corollary previously proved in [26].

Corollary 7.2.3 *Let $\Omega \subset \mathbb{R}^n$ and $\varphi(x, t) := t^p + a(x)t^q$ or $\varphi(x, t) := (t - 1)_+^p + a(x)(t - 1)_+^q$, $q > p > 1$. If $a \in C^\alpha(\Omega)$ is non-negative, then the maximal operator is bounded on $L^\varphi(\Omega)$ provided $\frac{q}{p} \leqslant 1 + \frac{\alpha}{n}$.*

Baroni, Colombo and Mingione [8] considered the border-line ($p = q$) double-phase functional $\varphi(x, t) := t^p + a(x)t^p \log(e + t)$.

Proposition 7.2.4 *The Φ-function $\varphi(x, t) = t^p + a(x)t^p \log(e + t)$ satisfies (A1) if and only if*

$$a(y) \lesssim a(x) + \frac{1}{\log(e + |x - y|^{-1})}$$

for every $x, y \in \Omega$.

In particular, if $a \in C^{\log}(\Omega)$, then φ satisfies (A1).

Proof We observe that $\varphi(x, t) \approx \max\{t^p, a(x)t^p \log(e + t)\}$. Hence we conclude that $\varphi^{-1}(x, t) \approx \min\{t^{1/p}, (t/a(x))^{1/p} \log(e + t/a(x))^{-1/p}\}$ and the (A1) condition becomes, after we divide by $t^{1/p}$,

$$\beta \min\left\{1, a(x)^{-1/p} \log\left(e + \tfrac{t}{a(x)}\right)^{-1/p}\right\} \leqslant \min\left\{1, a(y)^{-1/p} \log\left(e + \tfrac{t}{a(y)}\right)^{-1/p}\right\}$$

for $x, y \in B \cap \Omega$, $|B| \leqslant 1$ and $t \in [1, \frac{1}{|B|}]$. When we raise both sides to the power of $-p$ and discount the trivial case $1 \geqslant a(y) \log(e + t/a(y))$, we get the equivalent condition

$$\beta' a(y) \log\left(e + \tfrac{t}{a(y)}\right) \leqslant \max\left\{1, a(x) \log\left(e + \tfrac{t}{a(x)}\right)\right\}. \tag{7.2.2}$$

Since $a \log(e + \frac{t}{a})$ is increasing in a, the inequality is non-trivial only when $a(y) > a(x)$.

Denote $f(u) := u \log(e + \frac{1}{u})$ and $g(u) := u / \log(e + \frac{1}{u})$. Then

$$g(bf(u)) = \frac{bu \log(e + \frac{1}{u})}{\log(e + 1/(bu \log(e + \frac{1}{u})))} \approx u,$$

where the implicit constants depends on b. When we apply the increasing function g to both sides of (7.2.2) divided by t, we obtain the equivalent condition

$$\frac{a(y)}{t} \approx g(\beta' f(\tfrac{a(y)}{t})) \leqslant \max\left\{g(\tfrac{1}{t}), g(f(\tfrac{a(x)}{t}))\right\} \approx \max\left\{g(\tfrac{1}{t}), \tfrac{a(x)}{t}\right\}$$

or, equivalently,

$$\beta'' a(y) \leqslant t g(\tfrac{1}{t}) + a(x) = a(x) + (\log(e+t))^{-1}.$$

Taking into account that $t \leqslant \frac{1}{|B|} \leqslant \frac{c}{|x-y|^n}$, we can formulate the condition as follows:

$$a(y) \lesssim a(x) + (\log(e + |x-y|^{-1}))^{-1}$$

If $a \in C^{\log}(\Omega)$, then this certainly holds.   □

By Propositions 7.2.1 and 7.2.4 and Theorem 4.3.4 we obtain the following corollary.

Corollary 7.2.5 *Let $\Omega \subset \mathbb{R}^n$ and $\varphi(x, t) = t^p + a(x)t^p \log(e + t)$. If $a \in C^{\log}(\Omega)$ is non-negative, then the maximal operator is bounded on $L^\varphi(\Omega)$.*

We observe that the double phase functional satisfies (A2) with $h \equiv 0$. This means that we obtain modular inequalities without error term. For instance, from Proposition 6.3.12 we obtain the following version, which covers and improves on Theorem 1.6 in [26]. Similar results hold also for the other variants of the double phase functional.

Corollary 7.2.6 *Let $B \subset \mathbb{R}^n$ be a ball or a cube, $s \in [1, n')$, $1 < p < q$ and $a \in C^\alpha(B)$. Define $\varphi(x, t) = t^p + a(x)t^q$ and assume that $\frac{q}{p} \leqslant 1 + \frac{\alpha}{n}$.*

(a) *For every $u \in H_0^{1,1}(B)$ with $\varrho_\varphi(\nabla u) \leqslant 1$, we have*

$$\fint_B \varphi\left(x, \frac{|u|}{\operatorname{diam} B}\right) dx \lesssim \left(\fint_B \varphi(x, |\nabla u|)^{\frac{1}{s}} dx\right)^s.$$

(b) *For every $v \in W^{1,1}(B)$ with $\varrho_\varphi(\nabla u) \leqslant 1$, we have*

$$\fint_B \varphi\left(x, \frac{|u - u_B|}{\operatorname{diam} B}\right) dx \lesssim \left(\fint_B \varphi(x, |\nabla u|)^{\frac{1}{s}} dx\right)^s.$$

7.3 Other Conditions

In the papers [79–87, 94, 102–107, 123], Maeda, Mizuta, Ohno, Sawano and Shimomura considered Musielak–Orlicz spaces with six conditions on the Φ-function. They set

$$\varphi(x, t) := t\hat{\varphi}(x, t)$$

and study $L^\varphi(\Omega)$. Their conditions, with the notation of the present book, are:

(Φ1) $\hat{\varphi} \colon \Omega \times [0, \infty) \to [0, \infty)$ is measurable in the first variable and continuous in the second; and $\hat{\varphi}(0) = \lim_{t\to 0^+} \hat{\varphi}(t) = 0$, and $\lim_{t\to\infty} \hat{\varphi}(t) = \infty$.

(Φ2) $\hat{\varphi}(\cdot, 1) \approx 1$.

(Φ3) $\hat{\varphi}$ satisfies (aInc)$_\varepsilon$ for some $\varepsilon > 0$.

(Φ4) $\hat{\varphi}$ satisfies (aDec).

(Φ5) For every $\gamma > 0$, there exists a constant $B_\gamma \geqslant 1$ such that $\hat{\varphi}(x, t) \leqslant B_\gamma \hat{\varphi}(y, t)$ whenever $|x - y| \leqslant \gamma t^{-1/n}$ and $t \geqslant 1$.

(Φ6) There exist a function $g \in L^1(\mathbb{R}^n)$ and a constant $B_\infty \geqslant 1$ such that $0 \leqslant g(x) < 1$ for all $x \in \mathbb{R}^n$ and

$$B_\infty^{-1}\hat{\varphi}(x, t) \leqslant \hat{\varphi}(y, t) \leqslant B_\infty\hat{\varphi}(x, t)$$

whenever $|y| \geqslant |x|$ and $t \in [g(x), 1]$.

Note that the conditions vary slightly between their papers.

We observe first that there is no guarantee that the function φ above be increasing. In this sense the assumptions above are weaker than the framework adopted here. On the other hand, there are several restrictions compared to the assumptions in our present book. For now on we assume additionally that φ is increasing.

First, (Φ1) implies that φ is a Φ-prefunction of N-function type (see [34, Definition 2.3]):

$$\lim_{t\to 0} \frac{\varphi(x, t)}{t} = 0 \quad \text{and} \quad \lim_{t\to\infty} \frac{\varphi(x, t)}{t} = \infty.$$

This excludes L^1-type behavior. Second, (Φ2) implies (A0) in view of Corollary 3.7.5. The assumption (Φ3) implies that φ satisfies (aInc) (with exponent $1 + \varepsilon$, since $\varphi(x, t) = t\hat{\varphi}(x, t)$) and ($\Phi$4) implies that φ satisfies (aDec).

Condition (Φ5) looks quite similar to (A1$'$) in this book. Let us formulate both conditions in the format used in this book and highlight the differences (the latter condition is written for φ, not $\hat{\varphi}$, since they are equivalent):

(A1′) There exists $\beta \in (0, 1)$ such that

$$\varphi(x, \boxed{\beta} t) \leqslant \varphi(y, t)$$

for every $\boxed{\varphi(y, t)} \in [1, \frac{1}{|B|}]$, almost every $x, y \in B \cap \Omega$ and every ball B.

($\Phi5$) There exist $\gamma, \beta_\gamma \in (0, 1)$ such that

$$\boxed{\beta_\gamma} \varphi(x, t) \leqslant \varphi(y, t)$$

for every $\boxed{t} \in [1, \dfrac{\boxed{\gamma}}{|B|}]$, almost every $x, y \in B \cap \Omega$ and every ball B.

By (aInc)$_1$, we can always estimate $\varphi(x, \beta t) \lesssim \varphi(x, t)$, so in this sense the assumption in ($\Phi5$) is stronger than (A1′), unless also ($\Phi4$) holds, in which case it is equivalent (Lemma 2.1.10). Furthermore, if ($\Phi5$) holds for one γ_0, then it holds for other values as well: if $\gamma \leqslant \gamma_0$, then this is clear; if $\gamma > \gamma_0$, then we use doubling and almost increasing to handle $t \in (\frac{\gamma_0}{|B|}, \frac{\gamma}{|B|}]$. The essential difference is that (A1′) entails $\varphi(x, t)$ lies in $[1, \frac{1}{|B|}]$ whereas ($\Phi5$) entails that t lies in this range. Since φ satisfies (aInc)$_1$, the latter implies the former (the proof is analogous to Proposition 4.1.5).

The converse is not true. For instance, in Proposition 7.2.2, we showed that the double phase functional $\varphi(x, t) = t^p + a(x)t^q$ satisfies (A1) if and only if

$$a(y) \lesssim a(x) + |x - y|^{\frac{n}{p}(q-p)}.$$

A similar argument shows that it satisfies ($\Phi5$) if and only if

$$a(y) \lesssim a(x) + |x - y|^{n(q-p)}.$$

Since the exponent is larger in the latter condition, the latter assumption is stricter.

It remains to consider ($\Phi6$). Recall that we defined the limit Φ-function

$$\varphi_\infty^+(t) := \limsup_{|x| \to \infty} \varphi(x, t)$$

in Sect. 2.5. Assume first that ($\Phi1$), ($\Phi2$), ($\Phi3$) and ($\Phi6$) hold and φ is increasing. If $s, t \in [g(x), 1]$, then $\varphi(x, s) \leqslant B_\infty \varphi_\infty(s)$ by ($\Phi6$). If $t \in [0, g(x)]$, then $\varphi(x, t) \leqslant D\varphi(x, 1)t \leqslant DAg(x)$ by conditions ($\Phi3$) and ($\Phi2$). Hence

$$\varphi(x, t) \leqslant B_\infty \varphi_\infty^+(t) + DAg(x)$$

for all $t \in [0, 1]$. Similarly we establish that

$$\varphi_\infty^+(t) \leqslant B_\infty \varphi(y, t) + DAg(y).$$

Note that $g \in L^1(\mathbb{R}^n)$. Since $\varphi(x, t) \leqslant \varphi(x, 1) \leqslant c$ for $t \in [0, 1]$ by (Φ2), we may use $h(x) := \min\{DAg(x), c\} \in L^1(\mathbb{R}^n) \cap L^\infty(\mathbb{R}^n)$. Hence (A2') holds.

To show that the converse implication does not hold, we consider an example. Let $\varphi(x, t) := t^2 + tl(x)$, where $l \in L^1(\mathbb{R}^n)$ is a function with range $[0, 1]$ which equals 0 in the left half-space and 1 on a finite measure, open neighbourhood of the ray $(0, \infty e_1)$. Let $\varphi_\infty(t) := t + t^2$ and observe that

$$\varphi_\infty(t) - l(x) \leqslant \varphi(x, t) \leqslant \varphi_\infty(t)$$

for $t \in [0, 1]$, and so (A2') holds. Fix $B_\infty > 1$ and assume that

$$B_\infty^{-1}\varphi(x, t) \leqslant \varphi(y, t) \leqslant B_\infty \varphi(x, t)$$

whenever $|y| \geqslant |x|$ and $t \in [g(x), 1]$. Let $t_0 := \frac{1}{B_\infty - 1}$ be the positive solution of $t + t^2 = B_\infty t^2$. Then the inequality $t + t^2 \leqslant B_\infty t^2$ does not hold when $t \in (0, t_0)$. But if $y := Re_1$ and $x \in B(0, R)$ is in the left half-space, then $\varphi(y, t) = t + t^2$ and $\varphi(x, t) = t^2$. Since we assume that $\varphi(y, t) \leqslant B_\infty \varphi(x, t)$ when $t \in [g(x), 1]$, this implies that $g(x) \geqslant t_0 > 0$ in the left half-space, which contradicts $g \in L^1(\mathbb{R}^n)$. Thus (Φ6) does not hold.

Ahmida, Gwiazda, Skrzypczak (Chlebicka), Youssfi and Zatorska–Goldstein [4, 22, 48] studied differential equations in generalized Orlicz spaces without growth assumptions on the Φ-function. They do, however, assume that φ is an N-function. Moreover, they consider non-isotropic Φ-functions defined, i.e. the norm of the derivative is based on $\varphi(x, \nabla u)$, not $\varphi(x, |\nabla u|)$.

Here we consider their continuity condition on the Φ-function in the special case that φ is isotropic. They assume that, for all $\delta < \delta_0$ and all $t \geqslant 0$,

$$\varphi(x, t) \leqslant c\Big(1 + t^{-\frac{a}{\log(b\delta)}}\Big)(\varphi_{Q_\delta}^-)^{**}(t),$$

where Q_δ is a cube with side-length 2δ. Let us show that this condition and (A0) imply (Φ5), and hence also (A1).

Since φ satisfies (A0), Lemma 2.5.16 yields that $\varphi_{Q_\delta}^- \in \Phi_w(Q_\delta)$, and hence Proposition 2.4.5 implies that $(\varphi_{Q_\delta}^-)^{**} \simeq \varphi_{Q_\delta}^-$. If $t \in [1, 1/|Q_\delta|]$, then

$$t^{-\frac{a}{\log(b\delta)}} \leqslant c\delta^{-\frac{an}{\log(b\delta)}} \leqslant c.$$

Hence for the same range, $t \in [1, 1/|Q_\delta|]$, their assumption implies

$$\varphi(x, t) \lesssim \varphi_{Q_\delta}^-(t),$$

so (Φ5) holds.

To show that the implication is strict, we consider the double phase functional. We observed above that (Φ5) holds in this case provided $a \in C^{n(q-p)}$. The condition of Gwiazda, Skrzypczak and Zatorska–Goldstein becomes

$$t^p + a(x)t^q \leqslant c\left(1 + t^{-\frac{a}{\log(b\delta)}}\right)(t^p + a_{Q_\delta}^- t^q).$$

Consider a cube so small that $-\frac{a}{\log(b\delta)} < q - p$. Now if $a(x) > 0 = a_{Q_\delta}^-$, then the previous inequality cannot hold for large t, since the exponent on the left (that is q) is larger than the exponent on the right (that is $p - \frac{a}{\log(b\delta)}$). If no such cube exists, then $a_{Q_\delta}^- > c > 0$, which means that $a \approx 1$, in which case $\varphi(x, t) \approx t^q$. Hence the condition of Gwiazda, Skrzypczak and Zatorska–Goldstein holds for the double phase functional only in the trivial case. (In the variable exponent special case, it is equivalent to log-Hölder continuity.)

Finally, let us mention some variants of (A1) which are not studied in this book but appear in some articles [54, 57, 58, 61] dealing with regularity of minimizers with generalized Orlicz growth. The most used is condition (A1-n), which states that

$$\varphi(x, \beta t) \leqslant \varphi(y, t) \quad \text{when} \quad t \in [1, \tfrac{1}{|B|^{1/n}}].$$

Here the range of t is determined as if φ were the function t^n (this is the reason for the n in the name of the condition). This condition is related to the case of bounded minimizers. If φ satisfies (aDec)$_n$, then (A1-n) implies (A1), whereas if φ satisfies (aInc)$_n$, then (A1) implies (A1-n). We note that in the double phase case it corresponds to a different exponent of Hölder continuity of the weight a, namely $\alpha \geqslant q - p$ instead of $\alpha \geqslant n(\frac{q}{p} - 1)$ for (A1). Furthermore, in [61] some stronger versions of (A1) were considered in order to achieve higher regularity of minimizer.

7.4 Orlicz Spaces

Orlicz spaces form a somewhat different kind of special case. It is trivial to observe that any $\varphi \in \Phi_w$ independent of x satisfies (A0), (A1) and (A2). However, in this case we can say more, since (A1) holds not just for $[1, \frac{1}{|B|}]$ but on $[0, \infty)$ and similarly (A2) holds with $h \equiv 0$. This is significant especially in the case of inequalities in modular form, as it is the restriction in (A1) which leads to the assumption $\varrho_\varphi(|\nabla u|) \leqslant 1$ while the restriction in (A2) leads to the error terms. Let us therefore summarize here these results in the Orlicz case. Note in particular that these results can be applied to $\varphi^\pm$, when they are non-degenerate (Lemma 2.5.16). This has turned out to be useful when dealing with PDE with non-standard growth.

We remark that this kind of estimate was established in the special case that $\varphi \in \Phi_s$ is independent of x and finite in [66]. The following results follow from Propositions 6.2.10 and 6.3.12 and Corollary 6.3.13.

Corollary 7.4.1 (Poincaré Inequality in Modular Form) *Let Ω be a bounded domain and let $\varphi \in \Phi_w$.*

(a) *There exists $\beta > 0$ independent of Ω such that*

$$\int_\Omega \varphi\left(\frac{\beta|u|}{\text{diam}(\Omega)}\right) dx \leqslant \int_\Omega \varphi(|\nabla u|)\, dx,$$

for every $u \in H_0^{1,1}(\Omega)$.

(b) *If Ω is a John domain, then there exists $\beta > 0$ depending on the John constant such that*

$$\int_\Omega \varphi\left(\frac{\beta|u - u_\Omega|}{\text{diam}(\Omega)}\right) dx \leqslant \int_\Omega \varphi(|\nabla u|)\, dx,$$

for every $u \in W^{1,1}(\Omega)$.

Corollary 7.4.2 *Let $B \subset \mathbb{R}^n$ be a ball or a cube, $s \in [1, n')$ and $\varphi^{\frac{1}{s}} \in \Phi_w$. Then there exists a constant $\beta \in (0, 1]$ such that the following hold.*

(a) *For every $u \in H_0^{1,1}(B)$, we have*

$$\fint_B \varphi\left(\beta\frac{|u|}{\text{diam } B}\right) dx \leqslant \left(\fint_B \varphi(|\nabla u|)^{\frac{1}{s}}\, dx\right)^s.$$

(b) *For every $v \in W^{1,1}(B)$, we have*

$$\fint_B \varphi\left(\beta\frac{|u - u_B|}{\text{diam } B}\right) dx \leqslant \left(\fint_B \varphi(|\nabla u|)^{\frac{1}{s}}\, dx\right)^s.$$

Corollary 7.4.3 (Sobolev–Poincaré Inequality in Modular Form) *Let $B \subset \mathbb{R}^n$ be a ball or a cube, $\varphi \in \Phi_w$ and $s \in [1, n')$. Then there exists a constant $\beta \in (0, 1]$ such that the following hold.*

(a) *For every $u \in H_0^{1,1}(B)$, we have*

$$\left(\fint_B \varphi\left(\beta\frac{|u|}{\text{diam } B}\right)^s dx\right)^{\frac{1}{s}} \leqslant \fint_B \varphi(|\nabla u|)\, dx.$$

(b) *For every $v \in W^{1,1}(B)$, we have*

$$\left(\fint_B \varphi\left(\beta\frac{|u - u_B|}{\text{diam } B}\right)^s dx\right)^{\frac{1}{s}} \leqslant \fint_B \varphi(|\nabla u|)\, dx.$$

References

1. T. Adamowicz, O. Toivanen, Hölder continuity of quasiminimizers with nonstandard growth. Nonlinear Anal. **125**, 433–456 (2015)
2. R. Adams. *Sobolev Spaces*. Pure and Applied Mathematics, Vol. 65 (Academic Press [A subsidiary of Harcourt Brace Jovanovich, Publishers], New York-London, 1975)
3. R. Adams, J. Fournier, *Sobolev Spaces*, volume 140 of *Pure and Applied Mathematics (Amsterdam)*, 2nd edn. (Elsevier/Academic Press, Amsterdam, 2003)
4. Y. Ahmida, I. Chlebicka, P. Gwiazda, A. Youssfi, Gossez's approximation theorems in Musielak-Orlicz-Sobolev spaces. J. Funct. Anal. **275**(9), 2538–2571 (2018)
5. A. Almeida, P. Harjulehto, P. Hästö, T. Lukkari, Riesz and Wolff potentials and elliptic equations in variable exponent weak Lebesgue spaces. Ann. Mat. Pura Appl. (4) **194**(2), 405–424 (2015)
6. K.F. Andersen, R.T. John, Weighted inequalities for vector-valued maximal functions and singular integrals. Studia Math. **69**(1), 19–31 (1980/81)
7. P. Baroni, M. Colombo, G. Mingione, Harnack inequalities for double phase functionals. Nonlinear Anal. **121**, 206–222 (2015)
8. P. Baroni, M. Colombo, G. Mingione, Nonautonomous functionals, borderline cases and related function classes. Algebra i Analiz **27**(3), 6–50 (2015)
9. P. Baroni, M. Colombo, G. Mingione, Regularity for general functionals with double phase. Calc. Var. Partial Differ. Equ. **57**(2), Art. 62 (2018)
10. A.C. Barroso, E. Zappale, Relaxation for optimal design problems with non-standard growth. Appl. Math. Optim. (2018)
11. D. Baruah, P. Harjulehto, P. Hästö, Capacities in generalized Orlicz spaces. J. Funct. Spaces Art. ID 8459874, 10, (2018)
12. B. Bojarski, Remarks on Sobolev imbedding inequalities, in *Complex Analysis, Joensuu 1987*, volume 1351 of *Lecture Notes in Math.* (Springer, Berlin, 1988), pp. 52–68
13. S.-S. Byun, Y. Cho, J. Oh, Gradient estimates for double phase problems with irregular obstacles. Nonlinear Anal. **177**(part A), 169–185 (2018)
14. S.-S. Byun, J. Oh, Global gradient estimates for the borderline case of double phase problems with BMO coefficients in nonsmooth domains. J. Differ. Equ. **263**(2), 1643–1693 (2017)
15. S.-S. Byun, S. Ryu, P. Shin, Calderón-Zygmund estimates for ω-minimizers of double phase variational problems. Appl. Math. Lett. **86**, 256–263 (2018)
16. S.-S. Byun, Y. Youn, Riesz potential estimates for a class of double phase problems. J. Differ. Equ. **264**(2), 1263–1316 (2018)

© Springer Nature Switzerland AG 2019
P. Harjulehto, P. Hästö, *Orlicz Spaces and Generalized Orlicz Spaces*,
Lecture Notes in Mathematics 2236,
https://doi.org/10.1007/978-3-030-15100-3

17. C. Capone, D. Cruz-Uribe, A. Fiorenza, A modular variable Orlicz inequality for the local maximal operator. Georgian Math. J. **25**(2), 201–206 (2018)
18. R.E. Castillo, H. Rafeiro, *An Introductory Course in Lebesgue Spaces*. CMS Books in Mathematics/Ouvrages de Mathématiques de la SMC (Springer, Cham, 2016)
19. M. Cencelj, V.D. Rădulescu, D.D. Repovš, Double phase problems with variable growth. Nonlinear Anal. **177**(part A), 270–287 (2018)
20. V.V. Chistyakov, *Metric Modular Spaces*. SpringerBriefs in Mathematics (Springer, Cham, 2015). Theory and Applications
21. I. Chlebicka, A pocket guide to nonlinear differential equations in Musielak–Orlicz spaces. Nonlinear Anal. **175**, 1–27 (2018)
22. I. Chlebicka, P. Gwiazda, A. Zatorska-Goldstein, Well-posedness of parabolic equations in the non-reflexive and anisotropic Musielak-Orlicz spaces in the class of renormalized solutions. J. Differ. Equ. **265**(11), 5716–5766 (2018)
23. R. Coifman, C. Fefferman, Weighted norm inequalities for maximal functions and singular integrals. Studia Math. **51**, 241–250 (1974)
24. F. Colasuonno, M. Squassina, Eigenvalues for double phase variational integrals. Ann. Mat. Pura Appl. (4) **195**(6), 1917–1959 (2016)
25. M. Colombo, G. Mingione, Bounded minimisers of double phase variational integrals. Arch. Ration. Mech. Anal. **218**(1), 219–273 (2015)
26. M. Colombo, G. Mingione, Regularity for double phase variational problems. Arch. Ration. Mech. Anal. **215**(2), 443–496 (2015)
27. D. Cruz-Uribe, A. Fiorenza, *Variable Lebesgue Spaces: Foundations and Harmonic Analysis* (Birkhäuser, Basel, 2013)
28. D. Cruz-Uribe, A. Fiorenza, J.M. Martell, C. Pérez, The boundedness of classical operators on variable L^p spaces. Ann. Acad. Sci. Fenn. Math. **31**(1), 239–264 (2006)
29. D. Cruz-Uribe, P. Hästö, Extrapolation and interpolation in generalized Orlicz spaces. Trans. Am. Math. Soc. **370**(6), 4323–4349 (2018)
30. D. Cruz-Uribe, J.M. Martell, C. Pérez, *Weights, Extrapolation and the Theory of Rubio de Francia*, volume 215 of *Operator Theory: Advances and Applications* (Birkhäuser/Springer Basel AG, Basel, 2011)
31. D. Cruz-Uribe, L.-A.D. Wang, Extrapolation and weighted norm inequalities in the variable Lebesgue spaces. Trans. Am. Math. Soc. **369**(2), 1205–1235 (2017)
32. L. Diening, Maximal function on generalized Lebesgue spaces $L^{p(\cdot)}$. Math. Inequal. Appl. **7**, 245–253 (2004)
33. L. Diening, Maximal function on Orlicz–Musielak spaces and generalized Lebesgue spaces. Bull. Sci. Math. **129**, 657–700 (2005)
34. L. Diening, P. Harjulehto, P. Hästö, M. Růžička, *Lebesgue and Sobolev spaces with Variable Exponents*, volume 2017 of *Lecture Notes in Mathematics* (Springer, Heidelberg, 2011)
35. L. Diening, P. Hästö, A. Nekvinda, Open problems in variable exponent Lebesgue and Sobolev spaces, in *FSDONA04 Proceedings Drabek and Rakosnik (eds.); Milovy, Czech Republic* (Academy of Sciences of the Czech Republic, Prague, 2005), pp. 38–58
36. L. Diening, S. Schwarzacher, Global gradient estimates for the $p(\cdot)$-Laplacian. Nonlinear Anal. **106**, 70–85 (2014)
37. J. Duoandikoetxea, *Fourier Analysis*, volume 29 of *Graduate Studies in Mathematics* (American Mathematical Society, Providence, RI, 2001)
38. D.E. Edmunds, J. Lang, O. Méndez, *Differential Operators on Spaces of Variable Integrability* (World Scientific Publishing, Hackensack, NJ, 2014)
39. L. Esposito, F. Leonetti, G. Mingione, Sharp regularity for functionals with (p, q) growth. J. Differ. Equ. **204**(1), 5–55 (2004)
40. L. Evans, R. Gariepy, *Measure Theory and Fine Properties of Functions*. Studies in Advanced Mathematics (CRC Press, Boca Raton, FL, 1992)
41. X.-L. Fan, C.-X. Guan, Uniform convexity of Musielak–Orlicz–Sobolev spaces and applications. Nonlinear Anal. **73**, 163–175 (2010)

42. R. Ferreira, P. Hästö, A. Ribeiro, Characterization of generalized Orlicz spaces. Commun. Contemp. Math. (2019, to appear). https://doi.org/10.1142/S0219199718500797
43. F. Giannetti, A. Passarelli di Napoli, Regularity results for a new class of functionals with non-standard growth conditions. J. Differ. Equ. **254**(3), 1280–1305 (2013)
44. F. Giannetti, A. Passarelli di Napoli, M.A. Ragusa, A. Tachikawa, Partial regularity for minimizers of a class of non autonomous functionals with nonstandard growth. Calc. Var. Partial Differ. Equ. **56**(6), Art. 153, 29 (2017)
45. F. Giannetti, A. Passarelli di Napoli, A. Tachikawa, Partial regularity results for non-autonomous functionals with Φ-growth conditions. Ann. Mat. Pura Appl. (4) **196**(6), 2147–2165 (2017)
46. D. Gilbarg, N. Trudinger, *Elliptic Partial Differential Equations of Second Order*. Classics in Mathematics (Springer, Berlin, 2001). Reprint of the 1998 edition
47. R. Giova, Regularity results for non-autonomous functionals with $L \log L$-growth and Orlicz Sobolev coefficients. NoDEA Nonlinear Differ. Equ. Appl. **23**(6):Art. 64, 18 (2016)
48. P. Gwiazda, I. Skrzypczak, A. Zatorska-Goldstein, Existence of renormalized solutions to elliptic equation in Musielak-Orlicz space. J. Differ. Equ. **264**(1), 341–377 (2018)
49. J. Hadamard, Sur les fonctions entiéres. Bull. Soc. Math. France **24**, 186–187 (1896)
50. P.R. Halmos, *Measure Theory* (D. Van Nostrand Company, New York, NY, 1950)
51. E. Harboure, R.A. Macías, C. Segovia, Extrapolation results for classes of weights. Am. J. Math. **110**(3), 383–397 (1988)
52. P. Harjulehto, P. Hästö, The Riesz potential in generalized Orlicz spaces. Forum Math. **29**(1), 229–244 (2017)
53. P. Harjulehto, P. Hästö, Uniform convexity and associate spaces. Czech. Math. J. **68**(143)(4), 1011–1020 (2018)
54. P. Harjulehto, P. Hästö, Boundary regularity under generalized growth conditions. Z. Anal. Anwend. **38**(1), 73–96 (2019)
55. P. Harjulehto, P. Hästö, A. Karppinen, Local higher integrability of the gradient of a quasiminimizer under generalized Orlicz growth conditions. Nonlinear Anal. **177**(part B), 543–552 (2018)
56. P. Harjulehto, P. Hästö, R. Klén, Generalized Orlicz spaces and related PDE. Nonlinear Anal. **143**, 155–173 (2016)
57. P. Harjulehto, P. Hästö, M. Lee, Hölder continuity of quasiminimizers and ω-minimizers of functionals with generalized Orlicz growth. Preprint (2018)
58. P. Harjulehto, P. Hästö, O. Toivanen, Hölder regularity of quasiminimizers under generalized growth conditions. Calc. Var. Partial Differ. Equ. **56**(2), Art. 22 (2017)
59. P. Hästö, The maximal operator on generalized Orlicz spaces. J. Funct. Anal. **269**(12), 4038–4048 (2015)
60. P. Hästö, J. Ok, Calderón–Zygmund estimates in generalized Orlicz spaces. J. Differ. Equ. (2018, to appear). https://doi.org/10.1016/j.jde.2019.03.02
61. P. Hästö, J. Ok, Maximal regularity for local minimizers of non-autonomous functionals. Preprint (2019)
62. P.A. Hästö, Corrigendum to "The maximal operator on generalized Orlicz spaces" [J. Funct. Anal. **269**, 4038–4048 (2015)] [MR3418078]. J. Funct. Anal. **271**(1), 240–243 (2016)
63. J. Heinonen, T. Kilpeläinen, O. Martio, *Nonlinear Potential Theory of Degenerate Elliptic Equations* (Dover Publications, Mineola, NY, 2006). Unabridged republication of the 1993 original
64. H. Hudzik, Uniform convexity of Orlicz–Musielak spaces with Luxemburg norm. Comment. Math. Parce Mat. **23**, 21–32 (1983)
65. H. Hudzik, On some equivalent conditions in Orlicz–Musielak spaces. Comment. Math. Prace Mat. **24**, 57–64 (1984)
66. R. Hurri-Syrjänen, A generalization of an inequality of Bhattacharya and Leonetti. Can. Math. Bull. **39**(4), 438–447 (1996)
67. Y. Karagiorgos, N. Yannakakis, A Neumann problem involving the $p(x)$-Laplacian with $p = \infty$ in a subdomain. Adv. Calc. Var. **9**(1), 65–76 (2016)

68. Y. Karagiorgos, N. Yannakakis, A Neumann problem for the $p(x)$-Laplacian with $p = 1$ in a subdomain. J. Math. Anal. Appl. **454**(1), 412–428 (2017)
69. T. Karaman, Hardy operators on Musielak-Orlicz spaces. Forum Math. **30**(5), 1245–1254 (2018)
70. A. Karapetyants, S. Samko, On boundedness of Bergman projection operators in Banach spaces of holomorphic functions in half-plane and harmonic functions in half-space. J. Math. Sci. (N.Y.) **226**(4, Problems in mathematical analysis. No. 89 (Russian)), 344–354 (2017)
71. M.A. Khamsi, W.M. Kozlowski, *Fixed Point Theory in Modular Function Spaces* (Birkhäuser/Springer, Cham, 2015). With a foreword by W. A. Kirk
72. V. Kokilashvili, A. Meskhi, H. Rafeiro, S. Samko, *Integral Operators in Non-standard Function Spaces. Vol. 1*, volume 248 of *Operator Theory: Advances and Applications* (Birkhäuser/Springer, Cham, 2016). Variable exponent Lebesgue and amalgam spaces
73. V. Kokilashvili, A. Meskhi, H. Rafeiro, S. Samko, *Integral Operators in Non-standard Function Spaces. Vol. 2*, volume 249 of *Operator Theory: Advances and Applications* (Birkhäuser/Springer, Cham, 2016). Variable exponent Hölder, Morrey-Campanato and grand spaces
74. W.M. Kozlowski, *Modular Function Spaces*, volume 122 of *Monographs and Textbooks in Pure and Applied Mathematics* (Marcel Dekker, Inc., New York, 1988)
75. M.A. Krasnosel'skiĭ, J.B. Rutickiĭ, *Convex Functions and Orlicz Spaces*. Translated from the first Russian edition by Leo F. Boron. P. (Noordhoff Ltd., Groningen, 1961)
76. J. Lang, O. Méndez, *Analysis on Function Spaces of Musielak-Orlicz Type*. Monographs and Research Notes in Mathematics (Chapman & Hall/CRC, 2019)
77. A. Lerner, Some remarks on the Hardy–Littlewood maximal function on variable L^p spaces. Math. Z. **251**, 509–521 (2005)
78. D. Liu, J. Yao, A class of De Giorgi type and local boundedness. Topol. Methods Nonlinear Anal. **51**(2), 345–370 (2018)
79. F.-Y. Maeda, Y. Mizuta, T. Ohno, T. Shimomura, Approximate identities and Young type inequalities in Musielak-Orlicz spaces. Czech. Math. J. **63**(138)(4), 933–948 (2013)
80. F.-Y. Maeda, Y. Mizuta, T. Ohno, T. Shimomura, Boundedness of maximal operators and Sobolev's inequality on Musielak-Orlicz-Morrey spaces. Bull. Sci. Math. **137**(1), 76–96 (2013)
81. F.-Y. Maeda, Y. Mizuta, T. Ohno, T. Shimomura, Trudinger's inequality and continuity of potentials on Musielak-Orlicz-Morrey spaces. Potential Anal. **38**(2), 515–535 (2013)
82. F.-Y. Maeda, Y. Mizuta, T. Ohno, T. Shimomura, Hardy's inequality in Musielak-Orlicz-Sobolev spaces. Hiroshima Math. J. **44**(2), 139–155 (2014)
83. F.-Y. Maeda, Y. Mizuta, T. Ohno, T. Shimomura, Sobolev and Trudinger type inequalities on grand Musielak-Orlicz-Morrey spaces. Ann. Acad. Sci. Fenn. Math. **40**(1), 403–426 (2015)
84. F.-Y. Maeda, Y. Mizuta, T. Ohno, T. Shimomura, Duality of non-homogeneous central Herz-Morrey-Musielak-Orlicz spaces. Potential Anal. **47**(4), 447–460 (2017)
85. F.-Y. Maeda, Y. Mizuta, T. Shimomura, Growth properties of Musielak-Orlicz integral means for Riesz potentials. Nonlinear Anal. **112**, 69–83 (2015)
86. F.-Y. Maeda, T. Ohno, T. Shimomura, Boundedness of the maximal operator on Musielak-Orlicz-Morrey spaces. Tohoku Math. J. (2) **69**(4), 483–495 (2017)
87. F.-Y. Maeda, Y. Sawano, T. Shimomura, Some norm inequalities in Musielak-Orlicz spaces. Ann. Acad. Sci. Fenn. Math. **41**(2), 721–744 (2016)
88. J. Malý, W.P. Ziemer, *Fine Regularity of Solutions of Elliptic Partial Differential Equations*, volume 51 of *Mathematical Surveys and Monographs* (American Mathematical Society, Providence, RI, 1997)
89. P. Marcellini, Regularity of minimizers of integrals of the calculus of variations with nonstandard growth conditions. Arch. Ration. Mech. Anal. **105**(3), 267–284 (1989)
90. P. Marcellini, Regularity and existence of solutions of elliptic equations with p, q-growth conditions. J. Differ. Equ. **50**, 1–30 (1991)
91. P. Mattila, *Geometry of Sets and Measures in Euclidean Spaces*, volume 44 of *Cambridge Studies in Advanced Mathematics* (Cambridge University Press, Cambridge, 1995)

92. Y. Mizuta, E. Nakai, T. Ohno, T. Shimomura, Hardy's inequality in Orlicz-Sobolev spaces of variable exponent. Hokkaido Math. J. **40**(2), 187–203 (2011)
93. Y. Mizuta, T. Ohno, T. Shimomura, Sobolev's inequalities and vanishing integrability for Riesz potentials of functions in the generalized Lebesgue space $L^{p(\cdot)}(\log L)^{q(\cdot)}$. J. Math. Anal. Appl. **345**(1), 70–85 (2008)
94. Y. Mizuta, T. Ohno, T. Shimomura, Sobolev inequalities for Musielak-Orlicz spaces. Manuscripta Math. **155**(1–2), 209–227 (2018)
95. Y. Mizuta, T. Shimomura, A trace inequality of Riesz potentials in variable exponent Orlicz spaces. Math. Nachr. **285**(11–12), 1466–1485 (2012)
96. J. Musielak, *Orlicz Spaces and Modular Spaces*, volume 1034 of *Lecture Notes in Mathematics* (Springer, Berlin, 1983)
97. J. Musielak, W. Orlicz, On modular spaces. Studia Math. **18**, 49–65 (1959)
98. H. Nakano, *Modulared Semi-Ordered Linear Spaces* (Maruzen Co. Ltd., Tokyo, 1950)
99. H. Nakano, *Topology of Linear Topological Spaces* (Maruzen Co. Ltd., Tokyo, 1951)
100. A. Nekvinda, Hardy-Littlewood maximal operator on $L^{p(x)}(\mathbb{R}^n)$. Math. Inequal. Appl. **7**, 255–266 (2004)
101. C.P. Niculescu, L.-E. Persson, *Convex Functions and Their Applications*, volume 23 of *CMS Books in Mathematics/Ouvrages de Mathématiques de la SMC* (Springer, New York, 2006). A contemporary approach
102. T. Ohno, T. Shimomura, Trudinger's inequality and continuity for Riesz potentials of functions in Musielak-Orlicz-Morrey spaces on metric measure spaces. Nonlinear Anal. **106**, 1–17 (2014)
103. T. Ohno, T. Shimomura, Trudinger's inequality for Riesz potentials of functions in Musielak-Orlicz spaces. Bull. Sci. Math. **138**(2), 225–235 (2014)
104. T. Ohno, T. Shimomura, Boundedness of maximal operators and Sobolev's inequality on non-homogeneous central Musielak-Orlicz-Morrey spaces. Mediterr. J. Math. **13**(5), 3341–3357 (2016)
105. T. Ohno, T. Shimomura, Musielak-Orlicz-Sobolev spaces with zero boundary values on metric measure spaces. Czech. Math. J. **66**(141)(2), 371–394 (2016)
106. T. Ohno, T. Shimomura, Trudinger's inequality and continuity for Riesz potentials of functions in grand Musielak-Orlicz-Morrey spaces over nondoubling metric measure spaces. Kyoto J. Math. **56**(3), 633–653 (2016)
107. T. Ohno, T. Shimomura, Boundary limits of monotone Sobolev functions in Musielak-Orlicz spaces on uniform domains in a metric space. Kyoto J. Math. **57**(1), 147–164 (2017)
108. J. Ok, Calderón-Zygmund estimates for a class of obstacle problems with nonstandard growth. NoDEA Nonlinear Differ. Equ. Appl. **23**(4), Art. 50, 21 (2016)
109. J. Ok, Partial continuity for a class of elliptic systems with non-standard growth. Electron. J. Differ. Equ. **Paper No. 323**, 24 (2016)
110. J. Ok, Regularity results for a class of obstacle problems with nonstandard growth. J. Math. Anal. Appl. **444**(2), 957–979 (2016)
111. J. Ok, Regularity of ω-minimizers for a class of functionals with non-standard growth. Calc. Var. Partial Differ. Equ. **56**(2), Art. 48, 31 (2017)
112. J. Ok, Partial Hölder regularity for elliptic systems with non-standard growth. J. Funct. Anal. **274**(3), 723–768 (2018)
113. W. Orlicz, Über konjugierte Exponentenfolgen. Studia Math. **3**, 200–211 (1931)
114. K. Perera, M. Squassina, Existence results for double-phase problems via Morse theory. Commun. Contemp. Math. **20**(2), 1750023, 14 (2018)
115. V.D. Rădulescu, D.D. Repovš, *Partial Differential Equations with Variable Exponents*. Monographs and Research Notes in Mathematics (CRC Press, Boca Raton, FL, 2015). Variational methods and qualitative analysis
116. H. Rafeiro, S. Samko, Maximal operator with rough kernel in variable Musielak-Morrey-Orlicz type spaces, variable Herz spaces and grand variable Lebesgue spaces. Integr. Equ. Oper. Theory **89**(1), 111–124 (2017)

117. M.M. Rao, Z.D. Ren, *Theory of Orlicz spaces*, volume 146 of *Monographs and Textbooks in Pure and Applied Mathematics* (Marcel Dekker Inc., New York, 1991)
118. Y.G. Reshetnyak, Integral representations of differentiable functions in domains with a nonsmooth boundary. Sibirsk. Mat. Zh. **21**(6), 108–116, 221 (1980)
119. D.D. Rădulescu, V.D. Repovš, Q. Zhang, Multiple solutions of double phase variational problems with variable exponent. Adv. Calc. Var. (2019, to appear). https://doi.org/10.1515/acv-2018-0003
120. J.L. Rubio de Francia, Factorization and extrapolation of weights. Bull. Am. Math. Soc. (N.S.) **7**(2), 393–395 (1982)
121. W. Rudin, *Real and Complex Analysis*, 3rd edn. (McGraw-Hill Book Co., New York, 1987)
122. W. Rudin, *Functional Analysis*, 2nd edn. (McGraw-Hill Book Co., New York, 1991)
123. Y. Sawano, T. Shimomura, Sobolev embeddings for Riesz potentials of functions in Musielak-Orlicz-Morrey spaces over non-doubling measure spaces. Integral Transforms Spec. Funct. **25**(12), 976–991 (2014)
124. E. Stein, *Singular Integrals and Differentiability Properties of Functions*. Princeton Mathematical Series, No. 30 (Princeton University Press, Princeton, NJ, 1970)
125. B. Wang, D. Liu, P. Zhao, Hölder continuity for nonlinear elliptic problem in Musielak–Orlicz–Sobolev space. J. Differ. Equ. **266**(8), 4835–4863 (2019)
126. D. Yang, Y. Liang, L.D. Ky, *Real-Variable Theory of Musielak-Orlicz Hardy Spaces*, volume 2182 of *Lecture Notes in Mathematics* (Springer, Cham, 2017)
127. S. Yang, D. Yang, W. Yuan, New characterizations of Musielak-Orlicz-Sobolev spaces via sharp ball averaging functions. Front. Math. China (2019)
128. Q. Zhang, V.D. Rădulescu, Double phase anisotropic variational problems and combined effects of reaction and absorption terms. J. Math. Pures Appl. (9) **118**, 159–203 (2018)
129. V.V. Zhikov, Averaging of functionals of the calculus of variations and elasticity theory. Math. USSR-Izv. **29**, 675–710, 877 (1987)
130. V.V. Zhikov, On Lavrentiev's phenomen. Rus. J. Math. Phys. **3**, 249–269 (1995)
131. W. Ziemer, *Weakly Differentiable Functions*, volume 120 of *Graduate Texts in Mathematics* (Springer, New York, 1989). Sobolev spaces and functions of bounded variation

Index

© Springer Nature Switzerland AG 2019
P. Harjulehto, P. Hästö, *Orlicz Spaces and Generalized Orlicz Spaces*,
Lecture Notes in Mathematics 2236,
https://doi.org/10.1007/978-3-030-15100-3

LECTURE NOTES IN MATHEMATICS

Editors in Chief: J.-M. Morel, B. Teissier;

Editorial Policy

1. Lecture Notes aim to report new developments in all areas of mathematics and their applications – quickly, informally and at a high level. Mathematical texts analysing new developments in modelling and numerical simulation are welcome.

   Manuscripts should be reasonably self-contained and rounded off. Thus they may, and often will, present not only results of the author but also related work by other people. They may be based on specialised lecture courses. Furthermore, the manuscripts should provide sufficient motivation, examples and applications. This clearly distinguishes Lecture Notes from journal articles or technical reports which normally are very concise. Articles intended for a journal but too long to be accepted by most journals, usually do not have this "lecture notes" character. For similar reasons it is unusual for doctoral theses to be accepted for the Lecture Notes series, though habilitation theses may be appropriate.

2. Besides monographs, multi-author manuscripts resulting from SUMMER SCHOOLS or similar INTENSIVE COURSES are welcome, provided their objective was held to present an active mathematical topic to an audience at the beginning or intermediate graduate level (a list of participants should be provided).

   The resulting manuscript should not be just a collection of course notes, but should require advance planning and coordination among the main lecturers. The subject matter should dictate the structure of the book. This structure should be motivated and explained in a scientific introduction, and the notation, references, index and formulation of results should be, if possible, unified by the editors. Each contribution should have an abstract and an introduction referring to the other contributions. In other words, more preparatory work must go into a multi-authored volume than simply assembling a disparate collection of papers, communicated at the event.

3. Manuscripts should be submitted either online at www.editorialmanager.com/lnm to Springer's mathematics editorial in Heidelberg, or electronically to one of the series editors. Authors should be aware that incomplete or insufficiently close-to-final manuscripts almost always result in longer refereeing times and nevertheless unclear referees' recommendations, making further refereeing of a final draft necessary. The strict minimum amount of material that will be considered should include a detailed outline describing the planned contents of each chapter, a bibliography and several sample chapters. Parallel submission of a manuscript to another publisher while under consideration for LNM is not acceptable and can lead to rejection.

4. In general, **monographs** will be sent out to at least 2 external referees for evaluation.

   A final decision to publish can be made only on the basis of the complete manuscript, however a refereeing process leading to a preliminary decision can be based on a pre-final or incomplete manuscript.

   Volume Editors of **multi-author works** are expected to arrange for the refereeing, to the usual scientific standards, of the individual contributions. If the resulting reports can be forwarded to the LNM Editorial Board, this is very helpful. If no reports are forwarded or if other questions remain unclear in respect of homogeneity etc, the series editors may wish to consult external referees for an overall evaluation of the volume.

5. Manuscripts should in general be submitted in English. Final manuscripts should contain at least 100 pages of mathematical text and should always include

   - a table of contents;
   - an informative introduction, with adequate motivation and perhaps some historical remarks: it should be accessible to a reader not intimately familiar with the topic treated;
   - a subject index: as a rule this is genuinely helpful for the reader.
   - For evaluation purposes, manuscripts should be submitted as pdf files.

6. Careful preparation of the manuscripts will help keep production time short besides ensuring satisfactory appearance of the finished book in print and online. After acceptance of the manuscript authors will be asked to prepare the final LaTeX source files (see LaTeX templates online: https://www.springer.com/gb/authors-editors/book-authors-editors/manuscriptpreparation/5636) plus the corresponding pdf- or zipped ps-file. The LaTeX source files are essential for producing the full-text online version of the book, see http://link.springer.com/bookseries/304 for the existing online volumes of LNM). The technical production of a Lecture Notes volume takes approximately 12 weeks. Additional instructions, if necessary, are available on request from lnm@springer.com.

7. Authors receive a total of 30 free copies of their volume and free access to their book on SpringerLink, but no royalties. They are entitled to a discount of 33.3 % on the price of Springer books purchased for their personal use, if ordering directly from Springer.

8. Commitment to publish is made by a *Publishing Agreement*; contributing authors of multiauthor books are requested to sign a *Consent to Publish form*. Springer-Verlag registers the copyright for each volume. Authors are free to reuse material contained in their LNM volumes in later publications: a brief written (or e-mail) request for formal permission is sufficient.

Addresses:
Professor Jean-Michel Morel, CMLA, École Normale Supérieure de Cachan, France
E-mail: moreljeanmichel@gmail.com

Professor Bernard Teissier, Equipe Géométrie et Dynamique,
Institut de Mathématiques de Jussieu – Paris Rive Gauche, Paris, France
E-mail: bernard.teissier@imj-prg.fr

Springer: Ute McCrory, Mathematics, Heidelberg, Germany,
E-mail: lnm@springer.com

Printed in the United States
By Bookmasters